How To CHOP TOPS

In 301 Photos

Tim Remus

Motorbooks International
Publishers & Wholesalers

First published in 1996 by Motorbooks International Publishers & Wholesalers, 729 Prospect Avenue, PO Box 1, Osceola, WI 54020 USA

Motorbooks International books are also available at discounts in bulk quantity for industrial or sales-promotional use. For details write to Special Sales Manager at the Publisher's address

Library of Congress Cataloging-in-Publication Data

Remus, Timothy.
    How to chop tops in 301 photos/Timothy Remus.
        p. cm.
    Includes index.
    ISBN 0-7603-0068-2 (pbk.: alk. paper)
    1. Automobiles—Bodies—Design and construction—Handbooks, manuals, etc. 2. Automobiles—Customizing—Handbooks, manuals, etc.
I. Title.
TL255.R45   1996
629.26—dc20   95-26549

On the front cover: Dan Hix's Ford deuce coupe put in service as an A/gas drag car before Hix chopped, channeled, and restored the car. *Tim Remus*

On the back cover: Above: Hammer welding seamlessly joins sheet metal, but is a painstaking process. Below: Immaculately finished seams that require a minimum of filler are key to a quality top chop.

Printed in the United States of America

Contents

Chapter One

Planning the Chop

Chopping a top is often considered an essential part of building a hot rod. While the decision to chop your top may be a simple one, the process, of chopping a top even in the case of a relatively simple car like a Model A coupe, is an enormous undertaking. There's plenty of room for error—not small ones, but errors large enough to totally ruin an otherwise great car.

Many of the shops I spoke with while putting this book together told horror stories of cars that someone else had chopped badly, cars that eventually came to their shop to get "straightened out." Straightening out a badly done chop isn't only hard on a person's pride, it's also hard on the pocketbook.

Which is not to say a novice can't do a good job of chopping a top. Bob Larsen had never chopped a top, but he felt strongly that his Dodge Charger needed to have its roof lowered, and he did it—mostly in one weekend. At last report, he was putting the finishing touches on the body.

When people do have trouble, it's usually because they broke one of the rules for top chopping. The rules are pretty much the same, no matter who you ask. Roger Rickey and Jim Petrykowski and Kurt Senescall and all the other professionals consulted for this book offer nearly the same lists of things that should and should not be done.

A well-planned chop results in a totally new look to the car that compliments and accentuates the original lines of this Mercury.

On most do lists is the need to plan the chop. How far, at what angle, and by what means are just a few of the questions you'll need to answer before you start. In determining how far to go with a top chop, most people go to a car show and look for cars they like that are similar to their own cars. Magazines and books offer additional study aids, as does your own photo file of favorite cars.

Remember that how much you chop the top should be part of the overall design for the car. You need to keep all the proportions correct, so the top chop and the car's height and the length of the hood will all work together.

Though most street rod builders will drop the top the same amount in the front as in the back, a little rake one way or the other can have a subtle and very pleasing effect. George and Sam Barris often dropped the back of the roof more than the front to enhance a classic tail-dragging look. At Boyd Coddington's Hot Rod Shop they often take the opposite approach, dropping the front more than the back for a more aggressive look more suitable for a hot rod coupe.

The Paper Chop

Chopping a photocopied picture of your car is a good way to experiment before cutting any sheet metal. Though some professionals skip the "chop it on paper" stage, this kind of planning is a good idea for novice top choppers. Chopping a photograph of your car will help you figure out

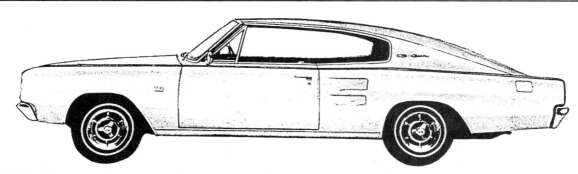

Before he started in with the saws and cut-off wheels, Bob Larsen took this line drawing of a Dodge Charger and chopped it with a scissors.

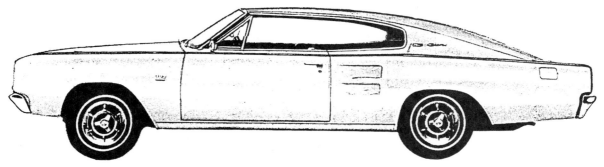

Dropping the top a modest amount and then slanting the windshield back to meet the new top position gave Bob the silhouette he was looking for.

how much to take out of the top, the benefits of a little rake one way or the other, and the impact of the chop on the car's overall looks. It will also help you anticipate problems that will occur when you do the real thing.

Mike Stivers of Car Creations in Blaine, Minnesota, is one professional who likes to start the planning process by chopping the car on paper. "I take a photo of the car from the side, an exact profile, then I go to a copy machine and blow it up as large as I can. Next I take that silhouette, get my scissors, and chop it on paper. Because if I can cut it on paper I can cut it in metal.

"I do that to get the appearance right and also to get customer approval. I can do the same thing with the computer," he added. "I scan in the photo and use a graphics program to chop it on the computer."

Once you have the right look on paper, you can compare known dimensions on the car with known dimensions on the photo to figure out the actual amount you've chopped the top. If the wheelbase, for example, is 100in in real life and 10in in the photo, the window that's 2in tall in the picture should be 20in tall in real life.

Building a car should be fun. By carefully planning the top chop, you minimize hassles and mistakes, maximize your money, get it done on time (maybe), and enhance the likelihood that the project will remain fun right up to the day you finish the paint job.

The Importance of Proportion

Doug Thompson, from Kansas City, Missouri, has a wealth of experience chopping tops. Though Doug is probably best known for his recent custom creations (Hirohata Merc clone, the *Radical 1946 Packard*, and the sectioned and shortened 1941 Ford, *String of Pearls*), he's also

Doug Thompson has been modifying cars for over forty years and has definite ideas about the best way to chop a top.

This 1950 Chevy is another Mike Stivers project. By dropping the top without adding any metal, and then leaning back the windshield, Mike was able to enhance the fastback lines of this Chevrolet.

Here you can see how the reveal around the windshield was trimmed back to enlarge the opening.

This 1934 Ford coupe with the 2.5in chop was seen at the Creative Metalworks shop. Kurt's idea was to keep all the windows the same vertical height.

The upper half of the rear window was removed before the top was dropped, and then reinstalled to maximize the size of the rear window.

built Model A street rods and radical customs from the 1960s.

Doug brings an extraordinary sense of balance to the business of top chopping. Though he understands the need for precision, he also understands the need to create a car with the right look. The word that came up again and again during the interview was "proportion." Better than most, Doug understands that cutting a top is more than just lowering the roof; it's a redesign of the entire car. So read along as Doug explains some of the things you should consider before you make that first cut.

Doug, how many tops have you chopped?

I'm not sure; I've been doing this stuff for forty-three years. In the early years, we sometimes did part of the job and then someone else would finish it, so it's hard to keep track. Most of those were simple jobs, nothing too fancy. No

Mike Stivers has a saying, "If I can cut it on paper, I can cut it in metal." Here we see the "before" photocopy.

The next step of a paper chop is to cut the top and the front and rear windows loose from the rest of the body.

Next, just drop the top straight down.

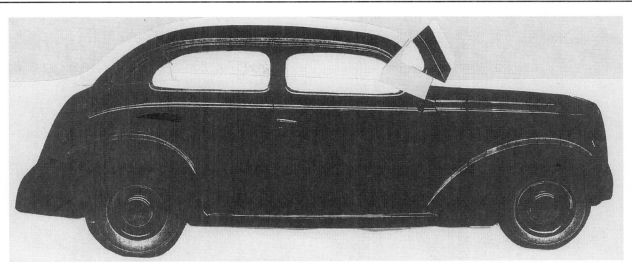

Lean the rear window in…

…then lean the windshield in and—presto—one chopped car. The beauty of this method is the opportunity it provides to try different chops and angles, and to anticipate problems.

real worry about proportioning. Over time, after it became boring to cut a top and drop it straight down, I progressed to wild art concepts. Maybe a new roof from a different car, different angles, and all the options that came to me at the time.

What are the planning steps you use?

It's hard to convey, because I don't get into a lot of detail in the planning stages. The way I work is to have a basic concept, a mental picture, but it may not have a lot of detail. It's like I can see this beautiful car in my mind, but the image is soft at the edges.

Sometimes I take a photo or photocopy of the car and I cut that up before I start on the car. I might determine how much of a cut I want to make by using a photograph, but after that I work mostly by eye. I'm very careful to make sure the top comes down the same amount on each side, but whether the cut is 2 or 3in isn't real important. "Proportion" is a more important word to me than "dimension."

One of your best-known cars is probably the Hirohata clone car built for Jack Walker. Can you walk us through that chop?

When I did the top on the Hirohata Merc, I knew basically what the Barris Brothers did,

9

In contrast to the five-window at Kurt's shop, this three-window at Boyd's shop was cut and dropped without cutting the top in two pieces. Note the metal work at the front posts, and the way the posts were leaned back to fit the new lower roof.

The upper door frame had to be cut into pieces and then carefully put back together to get the radius at the upper corner to match the roof. A 1982 Camaro roof was used to fill the top.

This 1941 Plymouth was chopped at Roger Rickey's shop. Metal work shows where the strip of steel was added to the roof to make up the length.

Roger used a second roof to extend the one on the 1948 by over 5in. Slanted posts were created by using post and upper door parts from the rear doors of a 1948 four-door. Rear window is Plexiglas and is glued in from the inside.

and I could tell that the roof tapered to the rear quite a bit. First, I determined how much of a chop I needed at the windshield posts—based on one of the old magazines that showed the original car.

Before I cut the top loose, I usually take the rear glass frame out. I reinforce the opening with braces so the area is well-supported. Then I cut out the window and a 2in flange all the way around. That's what I did on the Hirohata Merc.

Next, I cut the top loose and lower it until the steering wheel is in the right area of the windshield opening, as viewed from the front. That's where I tack-weld it. But I still have the whole tail end of the roof hanging loose.

So I make a pie-cut at the bottom of the front posts and start lowering the roof (leaning the windshield back at the same time). By looking at the side view of the car, and comparing that to the photos, I'm able to determine how far down to set the back of the roof.

People say the 1949 to 1951 Mercurys are the hardest to chop. Is that true?

The 1951 Mercury is harder for me than the two earlier models because of the larger rear window. It's hard to lay down the window and get a good side view of the car. Compared to a Model A coupe, a Mercury is a lot more work of course,

Next page
This 1948 Chevy (yes, that's right, it's a Chevy) is the favorite ride and personal pride of top chopper Roger Rickey. Roger chopped the top 5.5in at the front and 5in at the back. Note the gutterless roof and slanted door posts.

A study in contrasts. Note the chopped Deuce and the unaltered pickup behind it. *Rodder's Digest*

Another deuce, which goes to show that roadsters benefit from a lower lid just as much as coupes. *Rodder's Digest*

If you're going to slam the top and do all that metal work, why not create your own extended cab at the same time? *Rodder's Digest*

but a Merc isn't much harder than a 1950 Ford. I think the full-bodied cars with angled posts are harder. The Mercury is tough because there's the center post. It's a matter of the posts and at what angle do they lean; you've got more choices with the modern cars.

I don't split the roof into pieces when I do a Mercury. But lots of the younger guys are splitting the roofs. Guys who cut modern cars have trouble because the side windows lean in more than on the older cars, and when they cut the top, they lean in even more. And to cut one of those and not have the posts get too much angle, they have to cut up the roof and add material to the center. So sometimes when they do a Mercury roof they use the same approach: they either add metal, which makes for a bulky roof, or they re-skin the roof. Either way, it makes for a lot of work and gives a different visual effect than doing it the way I just described.

You said that you always check and double-check to make sure the roof is even from side to side and that everything is straight. Can you describe some of the techniques you use to do that?

I always use an "X measurement" so I know the top isn't leaning off to one side. I might mea-sure from the drip rail of the left side to the rock-er panel on the right side and compare that to the measurement from the drip rail on the right side to the rocker panel on the left side. When I'm working on the windshield opening I always do X measurements across the opening from one corner to the opposite corner to make sure they're even and square.

It gets complicated. You need to keep everything square and remember that you're working in three dimensions. The welding can draw the body out of square as you work. You need to check after each weld. If a guy gets in a hurry and does ten welds, then he checks and he's way off—well, then he's in real trouble.

You said that two people can get a different look from the same car, even with the same dimensions. Can you explain some of the options open to the top chopper that people might not consider?

Well, once the windshield frame is welded solid to the roof, then you can pie-cut either above or below the windshield and start moving the roof around at the tail end. I like to lean the windshield back; most of the early cars had a pretty severe straight-up-and-down angle on the windshield, and I like to change that angle.

15

We've been looking at these coupes with chopped tops for so long now, it's hard to imagine them without one. *Rodder's Digest*

I might drop the roof a lot, and then sometimes I move the windshield up into the roof, farther. On the Packard that I built for Richard Raty, I moved the windshield up into the roof so the windshield wouldn't get real narrow. It turned out that I cut four inches out of the windshield frame and six from the posts. So the windshield ended up two inches farther up into the roof than it used to be.

On a custom, I usually want the rear of the top to drop more than the front; rods are different, of course. I can usually make one cut and let the top down. The rear posts will slide down into the body cavity and then I can go and adjust it up and down until I get the angle I want from the side view. I mark the cut on one side and cut out the necessary piece and then use the piece I cut out to mark the same cut on the other side.

After the top is in position I can reinstall the rear window, either as it was or higher or lower—I like to consider all my options. Sometimes you need to move the original glass up or down, depending on what you've done with the rest of the roof. You can even use another rear window from a different car. What you do with the rear glass has a major impact on the car: you can shorten or lengthen the car visually; you can make a sedan look like a coupe, for example.

Talk about a lot of work—look at all that metal and all those door and window openings to keep straight and true while chopping the top on this 1948 Chevy. *Rodder's Digest*

How about some words of wisdom for the novice top chopper?

I can't over-stress the importance of proportioning, but early in the learning curve you don't want to do anything uneven or out of square. I've seen that happen. The irony is that a guy will get something crooked; he maybe sees it, but rather than go back and repair it, he fudges everything from that point on.

The people that I learned from taught me that as soon as you realize there's a mistake, you go back and repair it. For example, if you were building a house, you couldn't make the wall straight if the foundation wasn't level. You would need to go back and fix the foundation before going on.

Too many people will try to hide a mistake and go on. I believe you must correct it right away. The most important thing is to check your work frequently. We all make mistakes. You're going to make at least one somewhere. When you do, go back and correct it, and then go on. It seems costly to go back and make the correction at the time, but it's usually worse

Though street rods often benefit from having the front of the roof a little lower than the rear, customs like this Metranga-style coupe look natural with the back of the roof lower than the front. *Rodder's Digest*

This 1937 Ford sedan uses scallops to enhance the chopped top and lowered ride height, making the car seem even lower than it is.

This radical 1940 Chevy, once the property of Judy and Gary Lawrence, is chopped 5in. *Rodder's Digest*

to finish the job and find out then that it looks really bad.

I think most young guys are real hard on themselves. They think they need to be excellent right away, and when they make a mistake they think they're no good. They shouldn't be that hard on themselves. The true mark of your ability is how well you handle a challenge or a mistake. They need to remember that this car hobby is supposed to be fun; they shouldn't take it all too seriously. The job is never ruined by mistakes, as long as you correct each one before you proceed.

In addition to the chop, done by quartering the roof and adding material, the fenders are molded to the body, the headlights reside behind the grille, and the doors open in suicide fashion. *Rodder's Digest*

A specialized procedure requires specialized tools, and chopping a top is no different. The list of tools you need isn't terribly long, and with one exception, most aren't too expensive. First-time top choppers who tried to do the job with just a zinger and a hacksaw soon took the plunge and ran to the hardware store to buy more tools before finishing the job. What follows is a list of tools that most people need to do most top chops.

The Sawzall is a tough tool to beat when it comes to chopping a top. It is very nearly the only tool that reaches far enough to cut through both sides of a post or panel.

Sawzall Reciprocating Saw

You can cut a post with a zinger, you can cut it with a hacksaw, but nothing works like a Sawzall. With a long, fine-toothed blade, you can nearly always cut through both sides of a post or panel in one pass. They aren't exactly cheap, but when you consider how handy they are, it's hard to argue with the price. Besides the price, the only downside is the fact that the blade does walk around a little because it's controlled on only one side.

When you buy the saw, buy plenty of metal-cutting blades: long ones, for those times when you want to cut through both sides of a panel, and short ones, too, because there are times when you don't want to cut the other side. Buy extras because yes, they do break, especially on Sunday afternoon when none of the stores are open.

Die Grinder (or Zinger)

A small die grinder equipped with a cut-off wheel, and known on the street as a "zinger," is a great way to cut sheet metal and heavier braces

found inside of bodies. Though in most situations the wheel won't cut through both sides of a post or panel in one pass, there are still plenty of situations where a zinger is the tool of choice. If you don't already have one, buy a brand name tool, buy quality wheels (they last longer), and be sure to use safety glasses and hearing protection when you use the little work-saver.

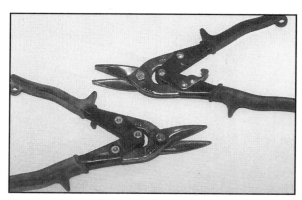

Be sure to equip your shop with a pair of good tin snips. They're inexpensive and a good way to cut and trim sheet metal without damaging the edge itself.

Tin Snips

Though they might seem rather low-tech, a high-quality tin snip is easy to control and leaves a clean edge, with very little distortion of the metal. When a cut needs to be precise, like when you're cutting a patch to fit in a hole, it's easy to make the rough cut with a zinger and then do the final cut with a tin snip.

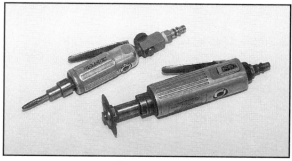

Known on the street as a "zinger," a small die grinder equipped with a cut-off wheel will cut through metal, both thin and heavy. Because the wheel is exposed, be sure to wear safety glasses and hearing protection—gloves are a good idea, too.

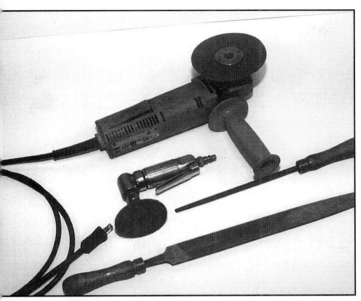

After you've made the cut, you need to trim the edges so that everything fits perfectly *before* you start welding. And the edge itself needs to be free of burrs prior to any welding. So you're going to need a variety of small grinders and files.

Grinders and Files

After making the cut, you need to trim the edge where the saw blade deviated from the line you carefully scribed on the metal. And burrs left by the zinger need to be removed before you start welding. Whether you use power tools or the files that Dad left behind, remember that sometimes it's nice to go slow and have total control. When you remove burrs, try not to bevel the edge of the sheet metal, which is already plenty thin!

Clamps, Clamps, and More Clamps

It's hard to have too many, so buy 'em by twos and threes. Buy long ones, short ones, fat ones, and needle-nosed ones—because you can't have the posts moving as you try to cut through the other side, and you don't want the top to fall off as you make the last cut on the last post.

Hammers and Dollies

You can't call yourself a metal worker or "tin man" without basic knowledge of how to work a hammer and dolly. For straightening

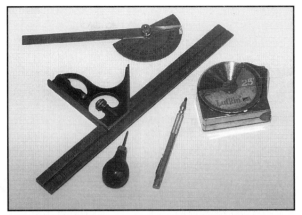

You can't cut until you've marked that line, and for that you need a simple square, scribe, and a good tape measure.

out the edge after cutting, and for flattening the seam as you weld it, you've got to have at least a couple of hammers and a couple of dollies. Buy hammers with different amounts of crown and a couple of dollies, also with different amounts of crown.

Scribes, Squares, and Tape Measures

Mom told you, your shop teacher told you, "measure twice, cut once." An experienced top chopper will spend more time laying out and scribing the lines than cutting and welding. Carpenter's tools are handy to have in a metal shop; sometimes it even helps to "snap

continued on page 22

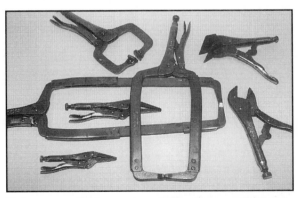

Fat ones, skinny ones, tall ones, and shorts ones. Ya gotta' love clamps because without them you will never get that top spliced and back together.

continued from page 21
a line." A little dark spray paint or bluing helps make the line you scribe easier to see. It's hard to have too much of this kind of stuff.

Welders

A heli-arc is the best welder to use for sheet metal work. The biggest advantage of a heli-arc welder is the small amount of heat used for a given weld, meaning less sheet metal warping. The biggest disadvantage is the high cost, followed by the intimidation factor. Some heli-arc welders are available at prices that seem somewhat reasonable, though. Miller, for example, makes a model called the "Econo TIG,"
which starts at less than $1,500.

Less expensive than a heli-arc is a gas-welding outfit or a wire-feed welder. Though purists complain that a wire-feed leaves a brittle weld and excessively warps the sheet metal near the weld, plenty of body work has been done with the durable little wire feed welders. For a 110-volt wire feed, you might try using .024in wire for sheet metal work, and for a 220-volt unit, .030in wire. If you've got to have a wire feed, buy a brand name and buy one with a higher-duty cycle.

Before anyone had a heli-arc, the best sheet metal welding was done with a gas-welding outfit. Not only can you weld with a

Perhaps the most important and least expensive tools in any shop—a pair of "ear muffs" and some good safety glasses.

Most professional shops rely on a heli-arc welder for welding sheet metal, as it's easier to control and uses less heat for a given weld, meaning less deformation of the delicate sheet metal.

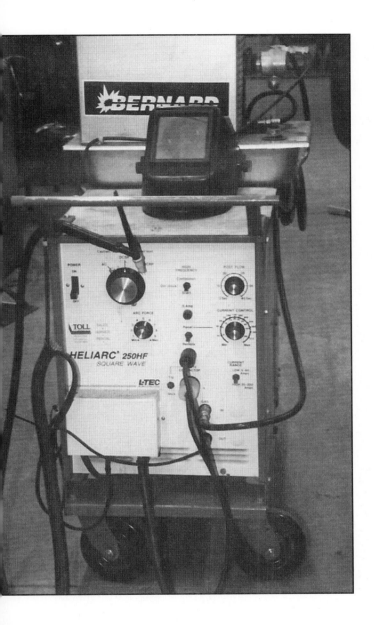

gas outfit, you can also heat, braze, and cut. Buy the best unit you can afford, and be sure to get tips small enough for sheet metal work. The tanks are usually leased; you probably don't need the big tanks commonly seen in commercial shops. Small tanks and a little cart of some kind will make it possible to roll the whole unit under the bench and out of the way when it's not in use.

Mini-Torch

When you weld together an old body, you're welding thin steel made thinner by years of rust and oxidation. Sometimes called a "mini-torch," an extra small torch for the gas-welding outfit makes it easier to weld that thin metal without burning holes in the edge.

If you don't already own any, buy at least two hammers (with different amounts of crown on the heads) and two dollies.

A mini-torch like this is very handy for welding sheet metal because it's so small and easy to hold, and also because it puts out less heat.

Here's the before picture of a 1951 Mercury being cut at Roger Rickey's shop. *Roger Rickey*

This photo shows how the rear window of a 1950 Mercury was reinforced and then installed in the 1951's roof. *Roger Rickey*

The braces allow Roger to move the window around without damaging the window frame. Note the slanted door post. *Roger Rickey*

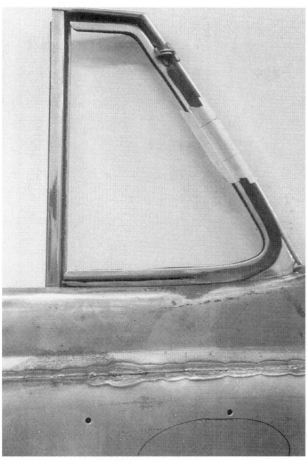

Planning includes figuring out what to do with windows and window frames. Steel or stainless steel frames can be cut and neatly welded. If the frames are cast, you probably have to find another way around the problem. Some builders like to eliminate the wing window entirely.

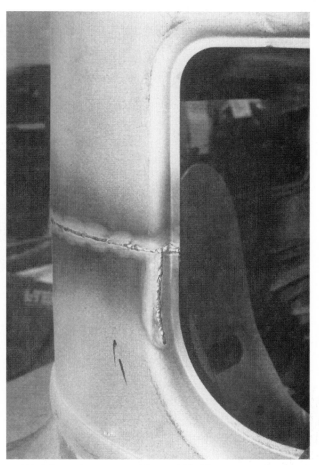

Welding sheet metal is a skill. It's easy to burn through the old metal bodies, or create so much heat that the door opening or top is drawn out of shape. If your own skills aren't up to par, tack-weld things together and then take the material to a skilled welder for the finish work.

Before you go looking for a roof section to fill the opening on your Model A, you need a template that shows the crown at the shoulder and across the top of the roof, so the new roof section will match up with the old.

Station wagon and van roofs work well for sedans. This roof came from a mid-1960s Mercury wagon.

Harry Ruthrauff chopped the top on this 1936 Ford a full 6in. Note how Harry got all the reveal lines perfectly aligned. *Rodder's Digest*

Although the tiny rear window on this chopped Ford is stylish, it's not especially functional. You should consider how much rearview mirror vision you are willing to sacrifice for the right look before you start cutting. *Rodder's Digest*

top-chopping tips

- Plan your cut carefully, making sure that the top chop enhances or complements the design of the entire car. When in doubt, find a photo or example of a top chop you like and copy it.
- Carefully measure all dimensions, including *all* window openings, before you start cutting.
- Carefully measure and mark the cuts.
- Try to make cuts at the top or bottom of a panel so that the cuts won't show and are easier to finish.
- Repair all body damage before cutting the top.
- Bolt the body to the frame with the frame sitting on a level floor.
- Reinforce the body so it won't open up after the top is cut off.
- If you're going to keep the wood framework around the windows, take the wood out first, trim it to fit the new dimension, and reinstall it after the top is cut.
- Don't stand in the car during the chopping process.
- Body or frame damage that you discover while cutting the top should be fixed before putting the top back together.
- After tack-welding everything back together, roll the car outside, step back from the car, and make sure you like the way it looks.
- Think of the body as the foundation of your house and the doors as trim. It's usually easier to fit the door to the opening, not the opening to the door.

Relatively square cars like this Deuce sedan are easier to chop than later, rounder designs. Note that the builder kept a relatively large windshield for a chopped sedan.

This side view shows the Hirohata clone car during construction. Doug built the car for Jack Walker, working entirely from old magazine photos. Note the raked windshield and missing center post.

The finished Hirohata car exhibits the long graceful lines characteristic of Barris brothers cars. In order to get those lines, they often dropped the back of the roof more than the front.

This construction photo shows the 1946 Packard that Doug Thompson built for Richard Raty. Creating the radical silhouette involved major surgery at the back of the body. The rear windows came from an earlier Dodge sedan.

Front view shows the relatively large front window area. Even though the front of the top was chopped 6in, the windshield was only cut 4in, meaning the windshield was actually moved up 2in into the roof.

Finished photo of the Packard shows the fade-away lines created by bringing the back of the top down much more than the front. In this case, the top chop was part of a total redesign.

Front view shows the relatively large windshield, slanted windshield posts, and a look that retains the Packard signature despite the extensive metal work.

Chapter Two

1926 Model T

This chapter documents a top chop done on a 1926 Model T at Creative Metalworks in Blaine, Minnesota. Most of the work was done by the owner of both the car and the shop, Kurt Senescall.

Kurt originally bought the car at an auction, for fifty dollars. At that time, the Model T was nothing more than a series of body panels hanging from the rafters of an old barn.

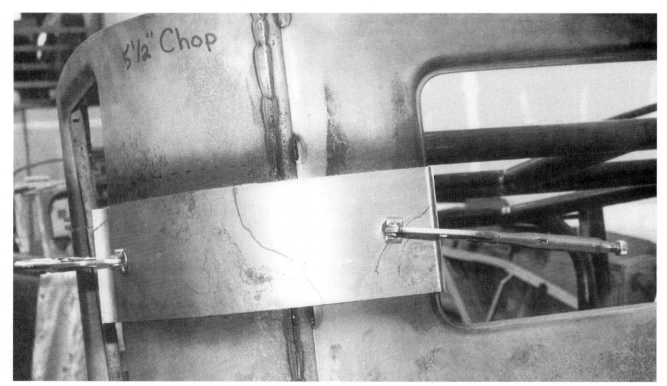

Cory's sketch and the rules dictate a 5.5in chop. Kurt cuts a piece of sheet metal exactly 5.5in across and uses that as a template to mark the cuts—making sure that it's level—before scribing the lines.

This Model T drag car project began as a $50 body and a paper sketch. The old Model T body is shown before the chop. Because the car is destined to be a race car, some of the finish work will lack the sophistication typically seen in a street rod.

Kurt decided to make a drag racer out of it, one with an old-time flavor. The initial planning for this car was a little unusual. Because Kurt wanted to race the car in National Hot Rod Association (NHRA)-sanctioned events, he had to meet the NHRA rules that set a minimum clearance between the driver's helmet and the roll cage. This forced Kurt to chop the top less than he might have otherwise.

Based on NHRA rules, designer Cory Harder drew a Model T that resembled a 1958 or 1960 drag car. That sketch served as the blueprint for Kurt's new car.

The chop is a little unusual in the sense that a drag race car isn't finished to the same degree as a street rod. Yet the various stages Kurt goes through in chopping the top, and the techniques themselves, apply to nearly any car.

The steps Kurt used to chop the Model T are the same he uses for most top chops:

1. Draw (or have drawn) a good sketch of the car.
2. Carefully mark where the top will be cut, scribing the lines you will follow with the saw or cut-off wheel.
3. Cut off the top.
4. Cut off the posts.
5. Set the top back on for a trial fit.
6. Trim the posts for a better fit.
7. Another trial fit.
8. Weld the top on, using a heli-arc welder and a minimum of filler.

Kurt likes to use butt-welds to reassemble the car, and emphasizes the importance of making sure everything fits perfectly before you begin welding the body back together. He will often cut just a little wide of the scribe line with the Sawzall and then use a small grinder to remove metal right to the scribe line.

Once everything fits just right, Kurt welds the top on with a heli-arc, trying whenever possible to do fuse-welds without the use of any filler rod. Though the doors are commonly fit to the car before the top is cut (and sometimes cut at the same time as the top), Kurt cuts the top first, then the doors, explaining that the old Model T body is pretty "flexible" and could be adjusted to meet the door, if necessary.

Because the Model T is essentially a race car, the finish work won't be done to quite the same degree as a fancy show car. Yet, most of the work necessary for a successful top chop is the same regardless of the intended use.

In Kurt's case, he started with a good sketch or plan before starting the layout process. The layout itself required nearly as much time as the actual cutting, because it's just as important. The cuts themselves were done very carefully, and then butt-welds were used to put everything back together again.

It's a case where haste makes waste—and a person hates to waste an old Model T body.

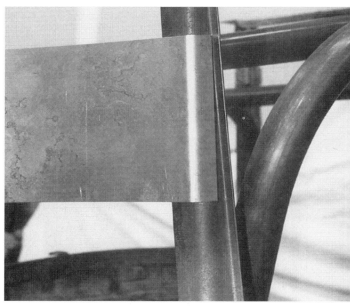

To mark the posts, Kurt simply bends the sheet metal so he can follow it around the corner.

The area to be marked is sprayed with black paint first so it will be easier to see the scribe line.

Kurt scribes the lines right through the rear window. If this were a street rod, he might want to keep the window bigger than a mail slot.

Here you can see the first cut along the upper scribe line. Kurt cuts the top off first, then takes out the center section, so he is always working on the body and not on the top sitting on the bench.

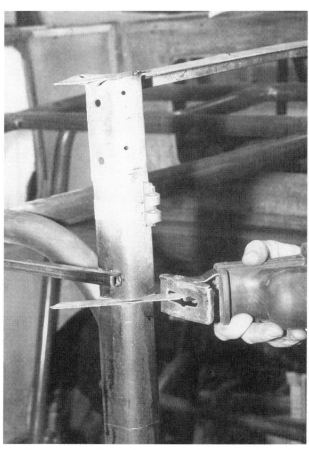

The cuts are done with a Sawzall, one equipped with a blade having either 18 or 24 teeth per inch.

This close-up shows that Kurt tends to cut a little inside the scribe line, then come back later and clean up the edge with a small sander.

Kurt and Bob Tinney lift off the top, which is really only a shell with no top panel.

With the top off, Kurt works his way around the car with the Sawzall. Kurt has Bob help support the sheet metal sections so he can do a nice, neat cut.

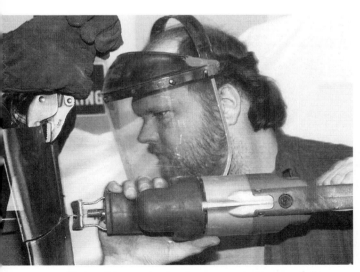

It's best to work slowly, concentrating on following that scribe line.

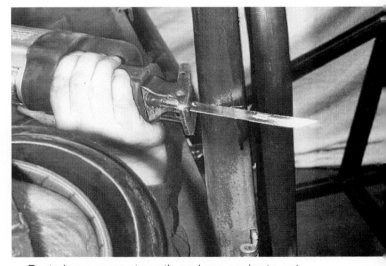

Posts have more strength and are easier to cut than large sections of sheet metal.

The 4in grinder equipped with a sanding pad is a good way to clean up the edges so everything fits as perfectly as possible.

The top is inverted and laid on the bench so the edges can be sanded right down to the scribe line.

The next step is a trial fit, followed by more sanding to get the two edges to meet without a lot of gaps. The top can then be clamped in place.

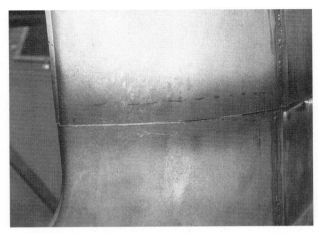

It's very important to do neat cuts so the body panels can be butt-welded together with little or no filler.

When the chopped top is attached to the body, spring clamps can be used for the initial positioning, followed by vise grips.

Kurt uses vise grips and a piece of square stock to keep the door posts lined up.

Most of the windows will be glued into this race car—so Kurt is careful to get all the inside window surfaces lined up correctly and flat.

Welding an area like this body corner is done by starting at one side and moving across to the other side, taking occasional breaks so everything will cool off.

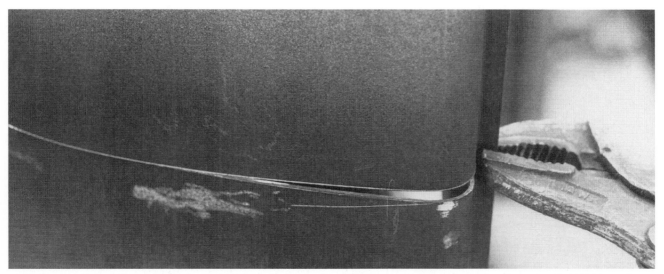

This shows how Kurt tacks both ends of the corner and then welds across from one end to the other. "I did it that way at the corners," he explains, "because I wanted any bulge to happen at one end of the weld next to the window. Then I can just slit one corner and eliminate the bulge, instead of having to deal with a corner full of lumps and puckers."

Here we see the Model T with most of the rear corners welded. All the welding is done by heli-arc; Kurt always tries to fuse-weld the sheet metal, which doesn't add any filler. In this case, the old T's body was so bad he used some oxyweld 65 rod (which does add filler), in either 1/16in diameter or in a .030in diameter (commonly found on a roll and used with a wire-feed welder).

This shot shows the bulge that's left on the driver's side after welding across the rear corner.

Both body and door posts have some taper to them; there's usually a mismatch where they join.

Kurt uses a variety of means to clamp and line up the window surfaces before tack-welding the posts.

Here, a piece of square tubing is used to line up the center post.

Once the post is lined up and either clamped or tack-welded, Kurt uses a cut-off wheel to make a pie-cut.

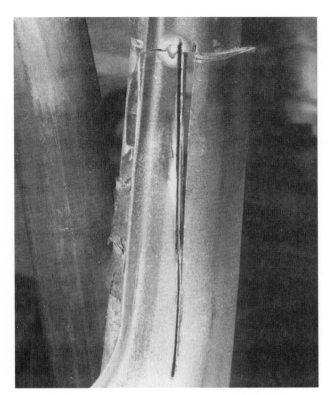

Here's the pie-cut. Kurt advises top choppers to make the cuts long so any movement occurs over a larger surface.

A large vise grip holds the channel in the right position while a straight edge is used to double-check the fit.

The channel is fit with a hammer and then heli-arc welded.

It's tough, but not impossible, to neatly weld on old thin body panels. The welds above are shown before being sanded.

Usually the doors are fit first, then cut at the same time the top is cut. But this is a race car with a flimsy body, so the doors are cut and fit after the top is chopped.

Getting the doors to fit requires pie-cuts on both sides of the post.

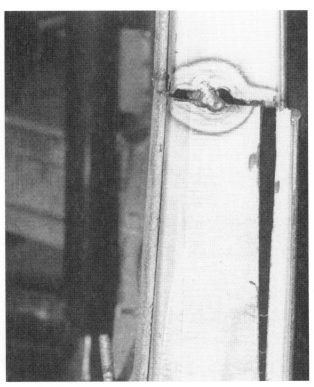

Here you see the pie-cut that will allow Kurt to move over the lower part of the door skin.

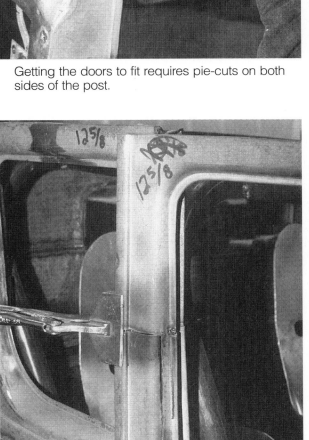

The outside skin must be pie-cut as well.

A little heli-arc welding later, the door is a one-piece unit.

This shows the same area on the other door after the initial sanding.

Nearly finished, Kurt uses square tubing and a hand-fabricated header panel to form the windshield opening.

Kurt's patient welding pays off in the nice, smooth sheet metal at the roof corner.

Next page
Here we see the finished (or partly finished) picture of the Model T. Not quite as low as Kurt would have liked, but it is tall enough to meet NHRA standards and lower and certainly more aerodynamic than the original.

Chapter Three

Model A

The top chop and roof filling operation seen here was done at Metal Fab in Minneapolis on a Model A belonging to Craig and Cindy Oldenburg. Craig decided on a 3.5in chop after careful study of some similar cars at shows and events.

The plan Craig and Jim Petrykowski laid out called for the top chop to be done more like a sec-tion job than a conventional top chop. Because the Model A is a pretty square car, the top could be dropped down intact.

Jim wanted to weld either high or low on the top, so the large panels wouldn't warp and so the welds would be easier to finish. Some of the welds could be hidden by body lines, and even those

Before starting on this Model A, Jim cuts the rear window, with a flange, completely out of the body.

Here we have the before picture of the Model A about to have its top hammered. Craig decided on a 3.5in cut after careful study of Model A Fords seen at shows.

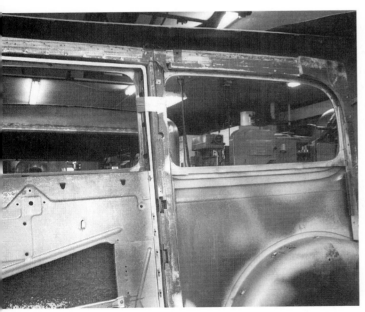

If the wood support structure is good, like it is in this deuce, it's tempting to just remove it before the cutting starts, then trim and reinstall it after the chop is finished. If the wood is gone, new wood kits can be ordered for many popular cars, or metal tubing can be substituted.

The layout of the cuts is just as important as the cutting itself. Before the cutting was begun, the window was removed and cuts were made past the body seam.

Jim carefully cuts along the scribed line with a cut-off tool. Because the Model A is a pretty square car, Jim designed the cuts to simply drop the top.

that couldn't would be close to a corner or reveal, which would make them easy to finish later.

The body was solidly bolted to the frame and reinforced before the top was cut off. Craig and Jim did the cutting with a Sawzall and a cut-off wheel. After the cut, the sections that were cut out of the posts were held in place with small vise grips in order to keep the top from collapsing at one corner.

The extra metal at the rear corners was left attached to the top and allowed to slide down inside the body when the top was set back onto the car. At this point, Jim and Craig did some careful measuring, making sure that the top dropped 3.5in at each corner and each post. This is where knowledge of the original body dimensions pays off.

Where the top slid past the body at the rear corners, the line where the top and body met was marked so the excess metal could be trimmed off. A sanding disc was used to trim the top exactly to the correct dimensions.

Cutting the posts at either the top or bottom eliminates the need to reconstruct the posts after they have been sectioned right in the middle. In order to make the top fit the new location, Jim slanted the front posts back slightly. The rear body corners were trimmed so the top would seamlessly butt to the body.

Where the front posts met the body, the base of the post (the part of the post still attached to the body) was slit at the corners so it could be tapered slightly to match the dimensions of the upper portion of the post.

The rear post (at the front of the rear window) was cut high, and when the top was dropped, pushing the top back slightly was enough to create a pretty good fit that wouldn't require a lot of finish work later.

Craig started work on the doors by removing 3.5in from the doors so that they matched the top posts. Rather than welding the cut-down door together and hanging it in the car with the assumption it would fit, Craig hung the bottom of the door and spent considerable time lining up the top part of the door.

Plenty of checking was done to ensure that the door fit the opening correctly before the final welding was done with the heli-arc.

Craig cuts the window posts with another air-powered cut-off tool. Note that all the cuts are made near the top or bottom of the opening—to make welding it all back together much easier.

Until they're ready for the big drop, all posts are temporarily clamped together.

The Sawzall is used in areas that can't be reached with the cut-off wheel. Note again that the cut is at the bottom of the post, not in the middle.

Liftoff. Now the easy part is done and the real work begins.

With the top off, it's easy to carefully cut some seam areas that will be butt-welded later.

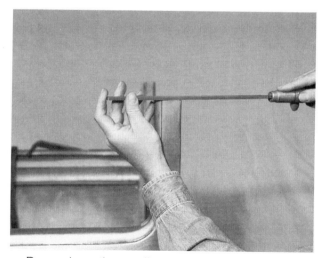

Burrs along the cut-line must be removed before welding everything back together. By keeping the file at a 90deg angle to the metal, burrs are removed without chamfering the already thin metal.

Time to begin putting the Model A back together again. If the layout and cutting were correct, there shouldn't be a need to take the top off again.

At this point, Jim and Craig have a Model A with two good doors and a chopped top, ready for the next step, filling the roof.

Fill a Roof

Jim Petrykowski says, "A filled roof is like an enormous patch panel—one that's located where it's easy to see but hard to install and hard to finish." The key difference between a roof and a regular patch panel is that you can't just go out and buy a roof. You've got to either have one made or find a suitable roof on an existing car.

Having one made can be expensive. Because most roofs need just a little crown, the filled section is often made by an experienced craftsman on an English wheel. Most hot rodders take the easier course and look for a roof that can be transplanted from a junkyard refugee to the coupe or sedan in the garage.

Before running out to the junkyard, you need to spend some time planning. The new roof needs to be in good condition, the right size, and have the right amount of crown before it will work. The first step is measuring; how big does the new panel need to be? If possible, the existing car's roof edges should be left intact. Most tops of old cars have more crown near the edge— if you cut the top way out near the edge, it makes it hard to match that crown to the crown of the transplanted roof.

Finding an appropriate roof section for a sedan is more difficult just because the area is so much bigger. Station wagons and vans are the likely donor cars, but you've got to find one with the style that you like (with or without ribs) and with the right amount of crown. Jim Petrykowski suggests that you look for a roof that's "on a car the junkyard has pretty well stripped—so it won't cost as much. Don't buy one that has any kind of obvious damage from people sitting on the roof, for example; damage to a roof is hard to repair, time-consuming, and expensive. Avoid wagon roofs with holes for luggage racks, and don't take one with any rust."

At the rear corners, the extra metal was left on the top, and it slid down past the outer body panel as the top was set back on.

Here's the corner after the extra metal has been trimmed from the top. All welds will be butt-welds, so everything must fit very nicely before the welding starts.

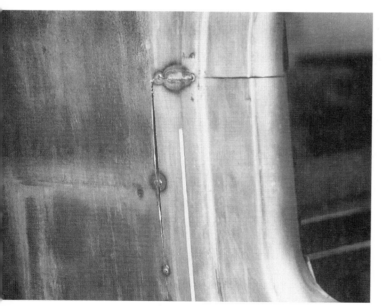

This tack-weld at the window post shows how well everything fits. After final welding, there won't be much finish work to do—and that's the whole idea.

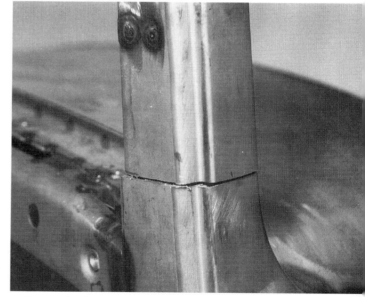

Here, the front post sits on the body. The new post is a little small for the base of the post, due to the taper, but it's not too bad. It's hard to see, but the base of the post has been slit at the corners and will be shrunk slightly to match the upper post.

The after picture. By planning carefully and then tilting the front posts back slightly, everything lines up without any major metal work.

When you tell the junkyard owner that you want the roof from that Mustang (good for coupes) or station wagon, be sure to take the entire roof, posts and all. Make sure it's cut off with a saw, so the headliner doesn't catch fire (and warp the roof). Also, be sure to take home the entire frame of the roof, rather than just the center skin, so that you don't damage your prize on the trip home.

Once you've got the whole roof home, clean it and sand it before cutting it free of its framework. Strip it of paint like you would any piece of metal and then get ready for plenty of planning and measurement time before the cutting and welding starts. When you do cut the new donor roof free of its supports, cut it with a zinger or shears so the top won't warp, and cut the piece as large as possible.

Before Craig went shopping for a roof, he and Jim made templates of the roof edges and the proposed roof center on the Model A so they would know what kind of crown the new roof should have. The crown of the roof is vitally important to your car for at least two reasons. First, the crown

of the donor roof must be such that it matches up with the shoulders of the existing roof. Second, how much crown the roof has through the center has a major impact on the overall look of the car.

Craig brought home the roof from a mid-1960s Mercury wagon. Once home, he stripped it of paint and then carefully cut it free of its framework.

The new roof was way too large for the Model A, and Jim spent considerable time moving it back and forth, trying to decide on the best natural fit.

When looking for the best natural fit, it helps to step back from the car, so you can actually see it. If the shop is small, it helps to tape the roof in what seems like the best position and then roll the car outside so you can stand back and better assess the fit, the crown, and the effect of the new top on the total look of the car.

Once Jim and Craig decided on the best fit, Jim trimmed overlapping metal at both ends of the top. Next, they carefully marked the center of the body and the center of the new roof, so it wouldn't be installed with an offset to one side.

what about the wood?

If you're cutting the top on an old coupe or sedan, you're going to have to decide what to do about the wood. Most cars from the 1920s and 1930s (and some later models) used wood around the inside of the window frames and across the top to give the body more strength. Before you chop the top, you need to decide how to deal with the wood. Some people automatically rip out the wood and replace it with a steel framework, but some of those builders act in haste. After all, the wood supported those old bodies for fifty or sixty years; there's no reason you can't keep the wood in your sedan for another five or ten.

If the wood in your car is in good shape, then just be sure to take it out before you cut the top. After the top is set back in place, carefully cut the vertical supports to the new dimensions and reinstall them around the windows. Don't cut the wood as you cut the metal and then try to patch the wood back together later.

It's when the wood is in bad shape that people assume they need to replace the old oak or hickory with metal framework. But before you go to all the work of fabricating new supports, consider the fact that replacement wood is available for many cars. Take a look through *Hemmings* or some of the restoration guides. Complete wooden support kits are available for most of the popular cars.

There is no one right way to replace that structure around the windows on your old hot rod. So just be sure to consider all your options before you assume that a new metal framework is the only way to go.

Above, **the wood on the inside of the '32 Sedan is in pretty good shape. In this case, it might be easier to just remove the wood, cut the top, and trim the wood to fit.**

Bottom right, **before you rule out wood supports as too old fashioned, consider the fact that many of the new fiberglass bodies use wood supports across the roof and around the doors.**

Here, Craig Oldenburg holds part of the tubular framework, fabricated at the Metal Fab shop, that will replace the original wood in his Model A (which was totally gone when he bought the car).

During this trial-fit period, Jim and Craig also determined the best angle and fit for the visor. Like the crown of the roof, the angle of the visor has a major impact on the overall look of the car. Some people put a header in across the top of the windshield before the roof is set on, but Jim and Craig decided the visor/roof assembly would provide so much strength that an extra header wasn't needed.

It's important that the crown at the edge of the roof match the crown at the edge of the opening. In some cases, a builder might have to cut the opening larger—though it's usually easier to do the welding and finishing on the flatter part of the roof, when possible.

Jim Petrykowski on Filling a Roof

After twelve years of chopping tops and filling roofs, and a number of years building race cars before that, Jim Petrykowski has a few guidelines for filling a roof. "Over the years we've probably had to redo at least forty roofs where the guy got it so screwed up he almost ruined the whole car. On most cars, it's the single biggest metal repair done to the car. If you measure the seam on the roof, it's a lot of welding and a lot of heat. I've seen cars where the welding caused the body to shrink so much that there were big gaps in the top of the doors—the top of the car actually shrank by an inch on either side.

"Usually, it's easier to lay the new roof over the opening in the car (after deciding where the roof fits best), scribe it from inside where the two edges meet, and then butt-weld the seam. If a guy overlaps the seam, welds it, and then finishes it later with filler, the seam will usually become visible on a hot summer day. Remember that the roof gets a lot of sunlight and that makes it real hot, especially on a dark car.

"I tell guys that, if they aren't real good at welding, just tack-weld the roof in place and then get someone real good to hammer-weld the seam, either with a heli-arc or a gas-welding outfit. Wire feed welders often make a hard seam that's hard to finish: it's brittle and sometimes cracks later.

"Whoever welds the top should move around, so they don't create a lot of heat in one area. Weld in one part of the roof and then move across the top and work in another area. Too much heat can shrink the whole car or it can shrink one corner so nothing ever fits right again. I try to avoid any seams in the center of the roof; they're real hard to finish later because the panel moves while you're working on it and you have to reach way out in the middle of the roof. Al-

Craig spends a lot of time test-fitting the upper and lower portions of the door, adjusting the hinges, and then fitting again before the two halves are welded together. Jim and Craig take their time; it's easier to make sure the door fits correctly now than it is later. The cuts are made at the bottom of the window opening, not at the middle.

After the initial welds, Jim does a bit of adjusting with a hammer and dolly.

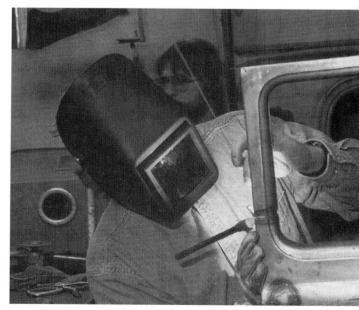

Inner welds are made last—after the upper and lower portions of the door fit correctly.

ways plan ahead, to minimize the amount of finish work needed—I try to make the body man's job as easy as it can be.

"Shrinking is a hard thing to do, especially for people who don't do it all the time. Guys at home should try to avoid situations where there's too much metal. If there gets to be a bulge at one part of the seam, I usually cut a little slit and try to flatten that area out as I go along so I don't have to go back and shrink it later. Anytime things start to go bad, maybe because the body isn't square or there's some old damage to the body that you didn't see before, stop and fix the damage. The problems are always harder to fix later.

"The top and body should be cut so the corners are rounded, because they're easier to weld that way. And the body should be well supported before you start filling the roof, so the heat from welding won't be so likely to pull it out of shape or shrink the top.

"Like I said, filling the roof is like doing a giant patch panel. There's a lot of welding and a lot of heat, all on an area that's hard to finish and easy to see when the car is done."

Jim decided this time to cut and weld the roof as he went along.

Before the welding started, Jim carefully cleaned the two edges, removing any burrs or raised metal. Burrs and nicks make it hard to do a neat weld and should always be cleaned off before the welding begins. The welding itself was done in a simple pattern. First, a three-inch-long section of the seam is tack-welded. Each tack-weld is hammered and dollied. After tack-welding, Jim carefully checks the area to see if the welding has created a bulge or concave area. If no corrective action is needed, then he welds the seam between two of the tack-welds, hammers and dollies that section, and then goes on to the next. All the work is done with a heli-arc welder.

A long seam like the roof is a time-consuming project. After working along one part of the seam for awhile, Jim either takes a break or moves to the other side of the car in order to avoid concentration of too much heat in one part of the roof.

By working the seam with a hammer and dolly, Jim is able to arrange the steel and filler molecules into a strong, forged pattern for a very strong weld. The hammer and dolly work also flattens and neatens the weld and minimizes the amount of warping at the welded seam.

Working in this slow, methodical fashion over the course of three days, Jim is able to finish welding the seam, and Craig now has a Model A with a filled roof. Jim decided not to grind or metal finish

A close-up of the finished weld. The door gap is even and the welds are neat and small. Note that the upper and lower parts of the front post have been blended together very nicely.

the roof for at least two reasons. First, large body grinders take too much metal, cause too much damage to the remaining metal, and often grind impurities from the pad into the metal. Jim explains that if he were going to grind the metal, he would either use the zinger (a die-cut grinder) and a cut-off wheel to carefully take the tops off the welds, or just take a grinder with a foam pad and an 80-grit pad and run that across the seam to remove the worst of the lumps and bumps.

In this case, Jim chose to do nothing, because, "Whatever kind of finishing you do should be done in concert with the person who's doing the body work. If I grind off the welds, the body man is probably going to grind it anyway and then we're taking off too much metal. Most of the time I leave the seam after I weld it and hammer and dolly it. That way the body man knows exactly what the seam or repair looks like and he can deal with the whole thing from start to finish."

The finished work. Jim and Craig decided early on that two hinges were plenty, and that it made the most sense to use the upper and lower ones, even though some builders use the lower and middle.

The top chop is finished, the doors are done, and the top is ready to be filled. How far out you trim the top opening depends on how good the metal is and what you have to do to get a good match between the crown of the opening and that of the new top.

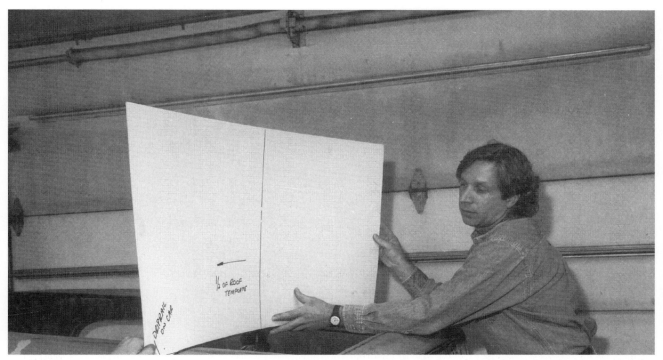

Before he went looking for a roof insert, Craig had a template like this one to take along, so he could find a roof with the right shape at the shoulder and the right amount of crown.

This is the new roof for Craig's Model A, donated by a mid-1960s Mercury station wagon.

As the roof is trimmed, there shouldn't be any big kinks or bows in it. The new roof should lay nice and flat on the car. From the inside, you shouldn't be able to see any big light leaks.

The wagon roof is way too big; Jim and Craig trim it after deciding which part of the roof has the shape they want. It's easy to put a roof section on a little to one side—Jim and Craig will carefully mark the center of the roof and line it up with the center of the car before determining the final position of the roof.

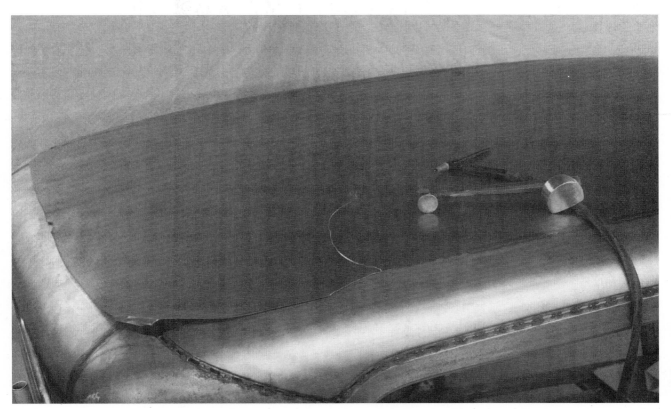

Here's the new roof, partly installed. Welding in a panel this big leaves plenty of room for error that can ruin the car. Novice welders might want to tack-weld the panel in place and let an experienced veteran weld up the seam.

The next few photos illustrate hammer welding. To begin with, trim the panel to fit. Though he sometimes cuts the insert to size before he starts welding, Jim decides to do the final trim with a set of high-quality tin snips as he works his way around the top.

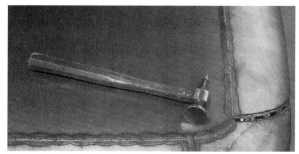

This close-up shows how neat the finished welded seam is—the result of all that painstaking work.

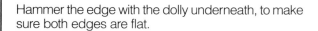

Hammer the edge with the dolly underneath, to make sure both edges are flat.

Tack-weld the flattened edges.

A file can be used to grind down a tack-weld that is too high.

Hammer and dolly the tack-weld. The hammer welding minimizes warping, flattens the weld, and also forges it, making the weld very strong. This process is repeated until the top is completely hammer-welded on. Jim moves around the car and doesn't work on one area for too long to minimize heat buildup and the resultant sheet metal warping.

Chapter Four

1934 Ford Coupe

This chapter takes a close look at the work-in-progress being performed on a 1934 Ford coupe at the Boyd Coddington shop. The man responsible for the work is one of Boyd's trusted craftsmen, Roy Schmidt. Roy has worked on everything from Duesenberg restorations to the finest hot rods and likes to do metal work that requires little or no body putty.

Though we only see part of this top chop, it's an important part. Roy is a master welder; he stitched the back of the roof so seamlessly that the metal warped imperceptibly and little body filler was required. Like other craftsman mentioned in this book, Roy prefers to avoid cutting the roof into two or more pieces.

Roy cut the top 2in at the front posts and 13/4in at the back of the door, to give the top a little rake. The

actual cuts were done near the top of the front post and at the bottom of the roof in back. The cuts themselves were done with a Sawzall; the top and doors were cut in one operation. It's interesting to note that at the back, Roy cut the top near

Side view shows the profile. Front posts were cut 1/4in more than the rear to give the top a little rake. Note that the front posts have been tilted back and the top front corner has been slit and opened up to stretch the top.

Rear view of this 1934 Ford coupe shows the top that was recently set in place. Note that the top was left in one piece and that the rear cut was done at the bottom of the top—except for the rear window area.

the bottom except for the rear window area, which was cut through the center. Roy explained that there is less slope near the bottom of the top, so there is less metal to make up. And by cutting through the center of the window opening, he avoids having to weld in a corner.

After cutting off the top, Roy set it back down, with the top sliding inside the body at the back (this is similar to the method used on the Metal Fab Model A in chapter three). After the top was clamped in position, a line was scribed where the top met the cut on the body. Then the top was pulled off and the metal trimmed along that line. To help make the top fit in the new position, the front posts were tilted back slightly and the front corner of the top was split and stretched to fit. After working on old cars for more years that he cares to admit, Roy swears this is much easier than cutting the roof into pieces or adding metal in the rear corners (which would involve much more cutting and, later, welding and finishing).

"Boyd cars" are known for having that certain look, it's like an attitude that no one can quite put their finger on. In reality, the look is achieved through careful planning and by keeping everything in proportion. In this case, the top wasn't cut too much, just enough. And by cutting the front posts just a little more than the rear, Roy puts a little rake on the roof. The rake is one of those subtle things that separate a car like this from all the others.

The front posts were cut near the top. The windshield post and side window posts have been slanted back slightly to make everything fit.

The inside of the coupe has been reinforced temporarily to keep everything in place when the top comes off. Note how the reinforcing forms natural triangles and also makes it easier to get into and out of the car, if necessary.

Here we see the results of the first series of tack-welds. Note how they are spaced evenly and how the gap is perfectly even where the two edges meet.

Roy chose to open up the front corner rather than slit the rear corner, which would have resulted in more seams, more filling, and more welding.

After he has welded an area, Roy comes back with the hammer and dolly, raising some areas and knocking others down. The constant hammer and dolly work creates a stronger, forged seam.

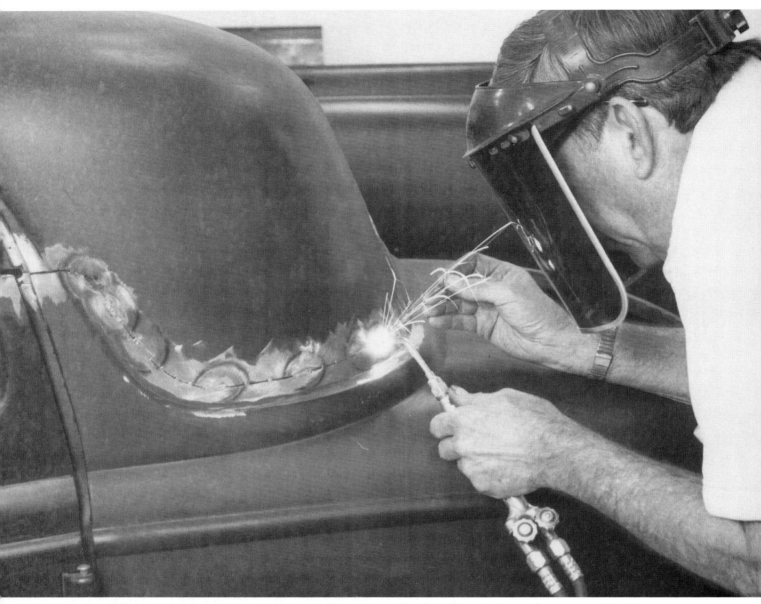

Roy follows a simple sequence when welding a seam like this. First, do just a few spot-welds. Then, come back and fill in with stronger, longer welds. Then, work the area with the hammer and dolly. Roy advises that the key is patience, "You've got to let everything cool off in order to avoid too much warping."

Roy uses a .030in rod for the filler and explains, "The small gap between the two pieces of metal will close up as you weld. If there isn't any gap, then the metal will create a crown as it is moved together by the heat of welding."

By keeping the cuts (and thus the welds) near the body seam, Roy avoids warping, because the seam is in a strong part of the body. Roy also likes to work near the seam because it acts as a heat sink and gives the heat created by welding somewhere to go.

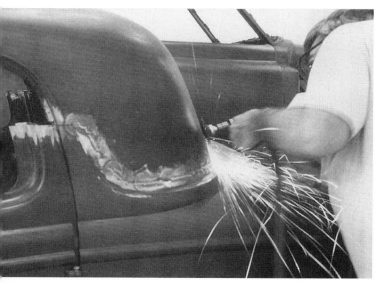

Occasionally, Roy knocks the high spots off the welds with a small grinder, so the dolly rests against the whole piece of metal and not just the top of a high spot.

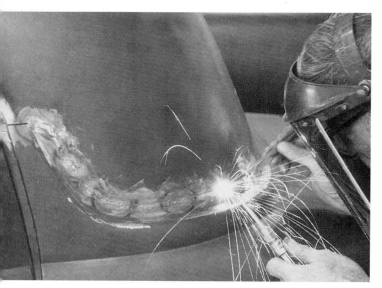

With a look of concentration just visible behind the mask, Roy finishes up the seam on the left rear of the coupe.

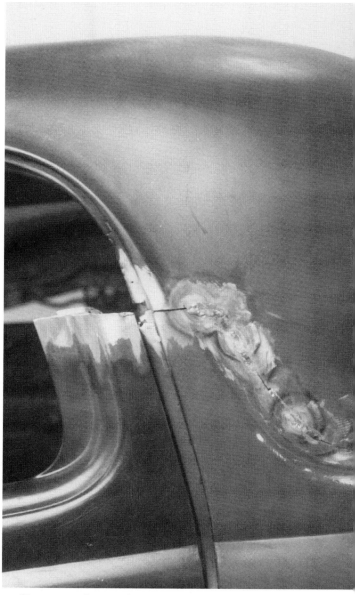

Close-up shows the very neat little welding beads and the way the gap is closing up even without any filler.

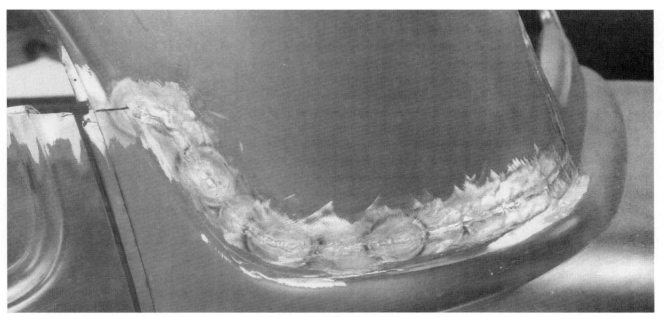

The further evolution of a welded seam. Note the even spacing of each welded area and the minimal warping.

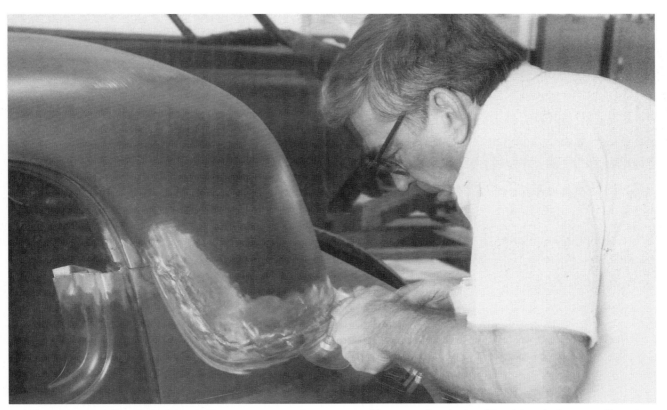

When the welding is finished, Roy goes over the seam with a small grinder. The idea is to smooth out the seam by eliminating the highest spots. Roy uses a small grinder to minimize gouging the metal and creating too much heat.

Here's a close-up of the seam before Roy is completely finished. Note that the area that actually needs finishing is just a narrow strip along the welded seam. One of Roy's goals is to minimize the need for body work later.

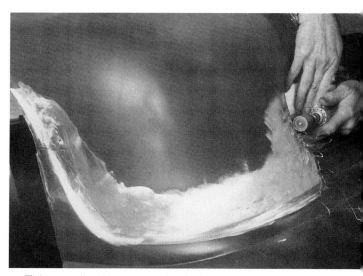

This small, air-powered grinder with a roll of sandpaper on the end is able to get into areas inaccessible to the other grinder.

Close-up shows a metal-finished seam. Even the raised area along the edge of the door is created with metal rather than built up with body filler.

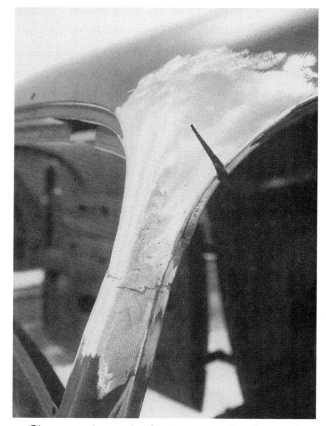

Close-up shows the front corner after the post is welded and before the slit is filled in.

The finished product. From here it's just a matter of using a little lead or body filler to finish the area and get it ready for primer.

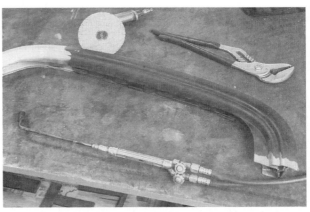

Because the front corner of the roof has been stretched, the radius at the top of the door must be altered as well. Here we see the top of the door and some tools Roy will use to transform the door top.

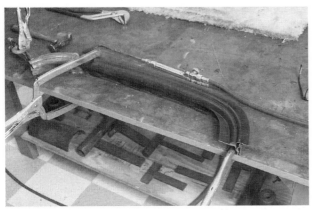

By clamping the door top to the bench, attaching a come-along to the front edge, and then heating the radius area, Roy will be able to alter the radius to match the top.

It isn't really pretty, but the setup is effective. After making his first "pull," Roy measures the door top.

This is the first trial fit of the door top, and the match is pretty close.

The door top changes width over its length, so the width is different here, where a section has been removed.

Side view of first fitting. The radius isn't quite perfect and needs just a bit more adjusting to work.

Here we see the results of some judicial slicing and tack-welding to get the inner edge of the two sections to line up. Roy likes to use the wire-feed welder for tack-welding because it's so quick and handy.

This close-up shows how Roy's patience has paid off in a door top that fits almost perfectly.

Roger Rickey has chopped a lot of tops. And he gets called in occasionally to act as advisor or consultant before, during, or after a chop. Roger advises people to ask for help before you start. There's no reason to make mistakes that can be avoided just by asking a few questions.

Roger recalls one situation where he was called in after the top was cut. "This one guy chopped a top and put it back together," explains Roger. "The car was a 1948 Chevy coupe. He cut the top way too low, and he didn't reinforce it ahead of time, so after he cut the top, the car caved in. The other thing he did was make way too many cuts around the back window. I never cut right above the back window like that.

"I don't know what happened to the car after I saw it, but if it were up to me, I would have gone out and found another top and just started over. People have to remember the basics. Always reinforce the inner body before you cut anything loose. Don't quarter the roof unless you really, really think you have to; it makes so much more work out of the job. And don't be afraid to ask for some advice, before you start on the job."

Almost finished; note the even door gap and matching radius at the front of the top. Roy will do the finish-welding with the door on the bench.

Chapter Five

1939 Plymouth Four-Door

This 1939 Plymouth is an unusual top chop, in part because it's a four-door, and also because it's done in what might be called a "one-top-chop" method. The shop is Car Creations in Blaine, Minnesota, owned and operated by Mike Stivers. The Plymouth is owned by Pat Barry of Lino Lakes, Minnesota.

Mike has been involved with old cars and car repairs for many years, and though he does everything from subframes to panel replacement, he likes chopping tops the best. Over the years he's developed his own way of doing tops and especially likes the one-top-chop. "When most guys chop a top, they use two tops," explains Mike. "They cut the top in two sections, front and back. Then they line up the windshield and back windows, then of course they need more material someplace in the middle of the roof. The two-top method adds 3 to 8in to the length of the roof and keeps the windshield and rear window at the stock angle. The one-top-chop (which I sometimes call the 'Bonneville chop') actually shortens the top three to eight inches, lays the windshield and back glass down more, and gives what I think is a more eye-appealing look."

Mike likes to start a top chop with a photograph. He takes a photograph of the car, preferably a side profile. He then uses a copy machine to enlarge the photo as much as possible. Next, he cuts up the photograph with a scissors and chops the top and reassembles the pieces until the car looks right.

Mike can also do this on a computer with graphics software and a scan of the photograph. This helps him plan for the chop. He also uses the finished paper chop to get customer approval.

For the Plymouth, Mike and the owner decided on what Mike calls, "a very practical 2.5in chop." Mike drops the top straight down, then leans the front and rear windshield back to meet the new top. Because the windshield isn't cut into a mail-slot, the occupants visibility is quite good.

This angle shows the windshield laid back to meet the roof—and the mismatch between the top and the metal surrounding the windshield.

This is where we start, with a 1939 Plymouth four-door at Car Creations in Blaine, Minnesota. At this point, the top has been dropped 2.5in, the windshield laid back (and only tack-welded in place), and the rear window/trunk area tipped forward. Major cuts were made across the roof (just behind the top of the windshield), through the door post, and just ahead of the top of the rear windows.

Mike recommends that top choppers reinforce the body, fit the doors, and make sure the body is bolted to the frame *before* starting any chopping.

On most cars, Mike can make cuts below the rear window or at the base of the trunk to lean the rear window forward. For the Plymouth, Mike pivoted the rear window and body section at the very base of the trunk. This works well and makes for an easy match at the seam—even the rain gutters line up this way. If you look at the seams on the back of the body, you will notice that Mike cut the upper part of the rear window frames out of the rear panel and trimmed the vertical length of the rear windows. This additional operation was done to improve the appearance of the Plymouth. Though he didn't do it on the Plymouth, Mike sometimes lowers the rear window in the rear body section to improve visibility.

Mike warns that leaning the rear window forward usually involves cutting the package tray so the rear of the body will come forward. On most cars, the trunk hinges are anchored under the package tray, so you have to keep the trunk, trunk lid, and the area where hinges are located as one unit so you don't change the geometry of the trunk and the way it opens and closes.

Mike cuts the windshield frame loose from the body by cutting the base of each post and then across the roof. Some slitting must be done at the base of the posts so they line up and have the right angle, but this method avoids having to cut the posts in the middle.

Cuts made across the top are made in the dead center of the crown, as viewed from the side. That way, when the two pieces are brought together they match up and have the same arc or curve. Sometimes Mike cuts a little off the center point at the front of the roof, depending on what kind of inner structure he is working around.

The dimensions of the car are carefully measured before any cutting is done, including vertical measurements all along the length of the roof.

Mike explains that "if you take out 2.5in you take it out consistently all through the car. Without those reference measurements, you're lost."

Mike always does a careful job of aligning the doors before starting on the top and then maintains that fit after the top is dropped. When the doors are cut and refit, the measurement must be made to the drip rail.

While most of the seams are butt-welded with the wire-feed welder, Mike stitch-welds the front roof seam and then goes in later and leads the seam. Mike prefers the stitch-weld on the front seam because he doesn't create as much heat as with a bead run all the way across the roof. By leading the seam he bonds the two pieces of metal together and ensures that the seam won't "witness" after the car is painted.

Though he works freely with lead, Mike only rough-finishes the seam with the material; the final work will be done with body putty applied over the lead. Any grinding is done with coarse-grit paper. The idea is to avoid turning the lead into dust that finds its way into your lungs.

If you choose to use lead, remember that it is a toxic material and exposure should be limited. Not just the lead but the fumes, too, are considered harmful, so a well-ventilated area is essential. Mike recommends you wear a dust mask and a hat to avoid breathing any dust and to keep the material out of your hair.

No matter how you finish the seams, the one-top-chop offers an alternative to more complex methods of top chopping.

Here you can see how the front posts have been cut so that the whole thing will lay back. By doing it this way you avoid having to find another top or weld strips of metal into the middle of the top. This method also keeps the windshield at stock size—so you don't need a periscope to see the stoplights.

Both doors are cut off at the same height. The rear door top has been tack-welded in place. Mike makes sure the doors fit correctly *before* starting on the top chop, and then works to maintain that correct fit.

Dropping the back of the top meant cutting through the rear-most post just under the hinge, through the window opening, and then up across the roof at the back.

Right side shows again how the top was cut and kept in one piece.

A close look shows the series of cuts that were made just above the rear quarter window, where the entire rear body panel was brought forward to meet the new top position. By "hinging" the body section at the very bottom, Mike was able to get the rain gutters to line up as the body section came forward.

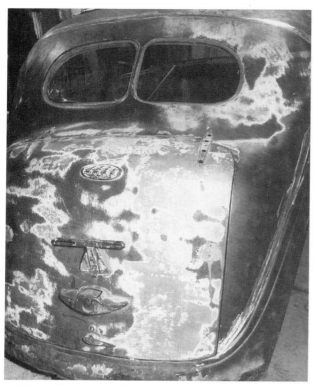

The entire rear section, including the trunk lid, was kept intact and pivoted forward, which also meant lots of work cutting and welding inner structures as well as the package shelf to make everything fit.

Here you can see the right side and how the top of the post area has been slit and moved over to meet the top.

Farther along in the project, the windshield and surrounding metal have been positioned. The left post is welded; right side is only stitched up.

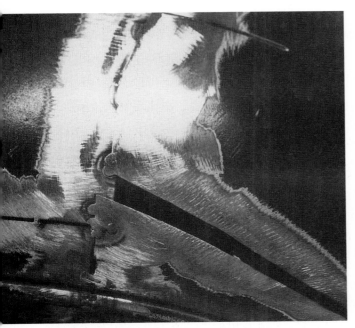

Close-up shows the large slit in the top of the post and another slit in the top itself. While Mike usually uses butt-welds, this top seam is overlapped slightly and will be leaded before he's done.

Left side shows the area all finish-welded. Center peak above windshield has been slit and flattened some to better match the contour of the top.

Now that the windshield section is in place, it's time to assemble the top of the door. Mike starts the job by cutting the top of the door into two pieces. The top of the door at the front is always one of the hardest pieces to rebuild.

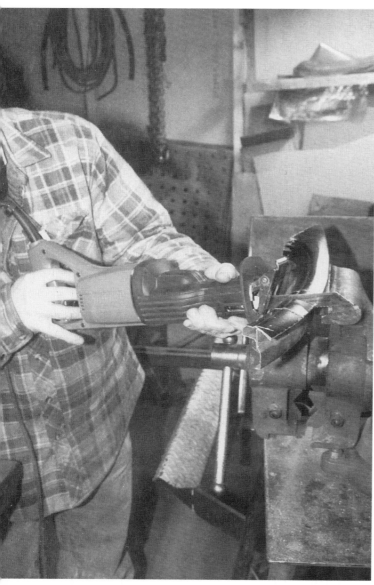

Then a trial fit.

Mike works methodically, first cutting a series of slots almost all the way through the new door section with a reciprocating saw.

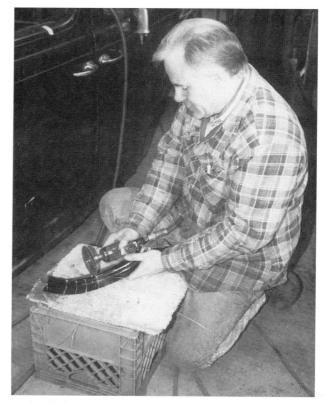

And then another slit to make it "flexible" enough to fit the contour of the top.

Of course, the width is never the same where the door meets the new door top, necessitating a series of slits in the inner door.

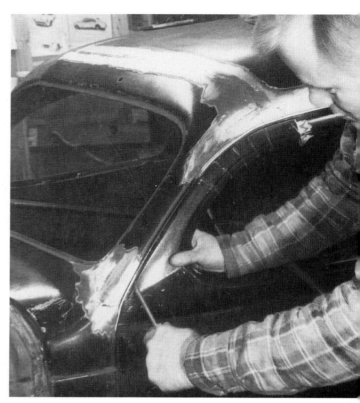

Here you can see the slits in the door top. After all that work, Mike must have a door that looks good from the outside, has a window channel that runs straight up and down, and provides enough clearance for a weather seal.

The new top section needs to fit correctly in three dimensions—and must match up to both the door and the windshield post.

Another trial fit shows that all those slits just might work—because the new section looks pretty good.

A small dip couldn't be eliminated along the top of the new door top, so Mike cut the new piece into two pieces, as shown, and will then cut a small piece to weld in.

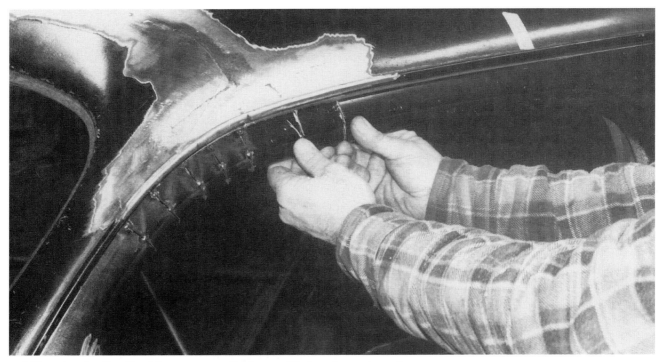

Mike checks the fit of the new small piece with the door closed.

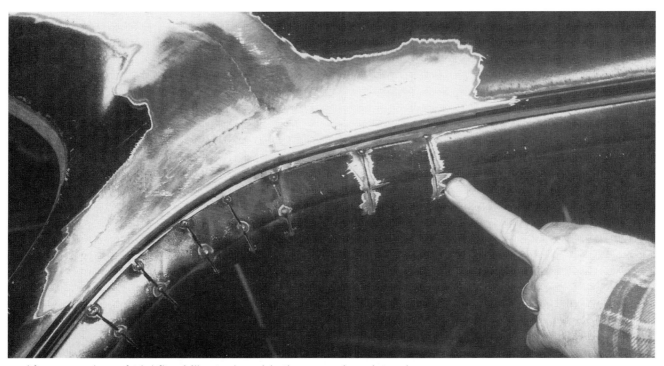

After a number of trial fits, Mike tack-welds the new piece into place.

The rough-finished door top with all the pieces in place.

Close-up shows the welds done with the wire-feed welder and how the door top does a good job of following the line of the rain gutter—Mike's primary goal in all the fittings.

Chris Pearson has been working at the Harmon Glass shop in suburban Minneapolis for over eleven years. The Roseville, Minnesota, shop where Chris works is one of those willing to take on unusual jobs, like classic cars, street rods, and glass for chopped tops. Before you chop that top, you'd better know exactly what you're going to do about the glass. And before you decide what to do, you need to listen to what Chris has to say about automotive glass.

Why don't we start by defining the basic types of automotive glass?

There's really only two, tempered and laminated. Laminated you can cut, tempered you cannot cut. Tempered glass is run through a kiln and heat-treated; you can't do anything with it.

How are the two types different?

Curved windshields are all laminated, so if you got hit by a rock, the rock won't shatter the glass. All the other glass in your modern cars is tempered. They do it that way so that, in an accident, the glass shatters, and you don't have a long piece of glass coming into the car, which could kill you. Some of the older hot rods have laminated glass in the door glass because you can cut it and it's cheaper.

So laminated glass is preferable for chopped cars or street rods?

Yes, and it comes in various colors: bronze, gray, tint, and clear.

Some people cut a template out of plywood and then bring that into the glass shop so the shop knows how big to cut the glass. Is that a good way to get glass cut to the right size?

Yes, that's OK, as long as it's a good pattern. The best way to go is to bring in the old glass and use that as the pattern. We would actually like to get the car into the shop so we can check the fit and grind the edge of the glass if we need to. But you've got some tolerance, probably an eighth of an inch either way.

People complain that when they have glass cut for the doors or whatever, sometimes it comes back with chipped edges or rough corners. Is that just sloppy work by the people who cut the glass?

Yes, the people who cut the glass just didn't take the time to grind the edges and smooth them off.

Street rodders who chop the top on a car with curved front and rear glass often have trouble cutting the windshield or having it cut; what do you recommend?

Well, if the windshield is laminated, we can cut it, but not very many glass shops will do that. [Note: At this point, Clem Brezinka, senior man in the shop, puts his head in the door of the lunchroom and offers additional advice: "The bigger the curve in the glass," he notes, "the more chance there is that it will crack when it's cut. The best way is to sandblast off the excess glass. The percentage of success is much higher that way than with grinding off the extra glass."]

What about the Polycarbonate or Lexan? What's the difference between the two and how do they work for front and rear glass?

[Clem again] They're just different brand names for essentially the same material. It's safe; you can hit it with a 20lb hammer and it won't break the material. You can cut it with a saber saw. It doesn't work well for a windshield because the wipers will scratch it, but it would work fine for back glass. I think that eventually it will yellow, because of the sunlight, but if the car is inside a lot, maybe that's not such a big deal.

Can this Polycarbonate material be shaped in an oven to form a curved glass?

[Back to Chris] Yes, with the right equipment you probably can. You can also have a piece of tempered glass made by a big glass company, but there's a big charge to make the form.

What are the mistakes people make when they cut glass for old cars?

They start with a bad pattern or no pattern at all. And they don't research the glass to figure out ahead of time what they're going to do.

The newer cars have tempered side glass, too. After about what year did they start using tempered in the side glass?

In about the early 1960s, all that side glass became tempered. And of course some of that glass is curved, which means it can't be cut and it can't just be replaced with laminated glass either.

Any final words of wisdom for street rodders who need custom glass?

Find a good glass shop that will work with you. And paint the car first, so there isn't any tape line after the glass is installed.

Chris Pearson has eleven years of experience working with automotive glass, and recommends laminated glass as the preferred windshield material for car builders.

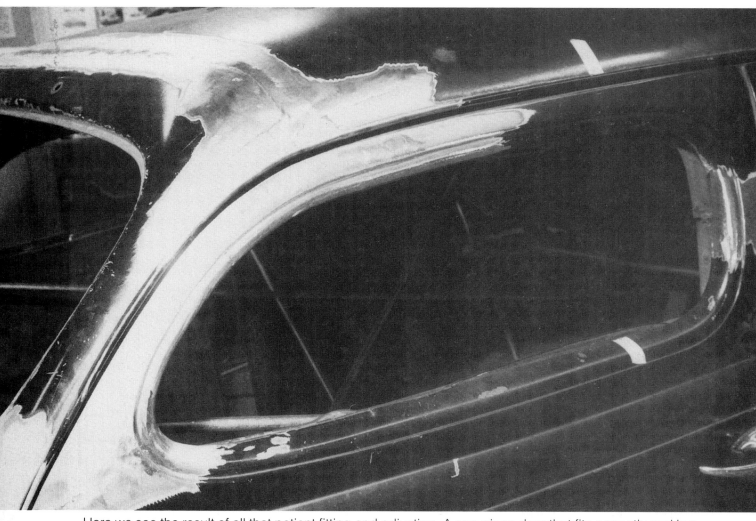

Here we see the result of all that patient fitting and adjusting. A one-piece door that fits correctly and has a nice even gap between the top of the door and the rain gutter.

Though the method of creating a new door top works well, this picture of the inside of the door top shows how it also creates the need for plenty of careful welding later.

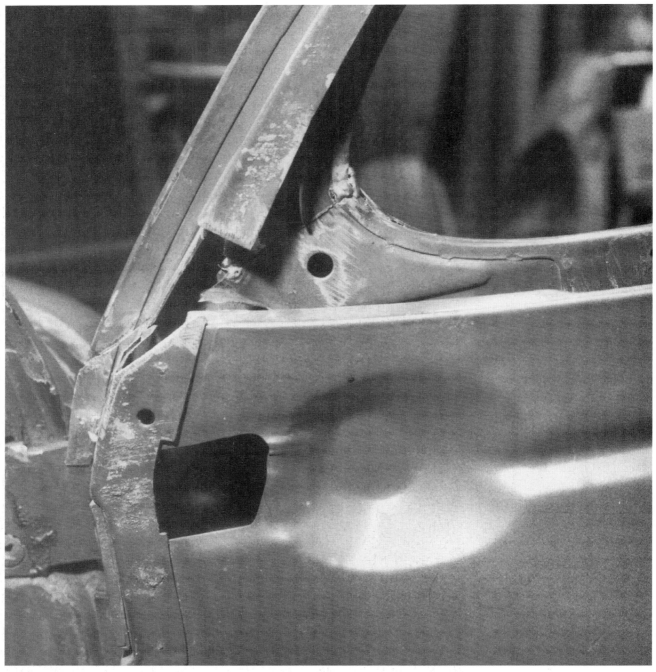

Making the door top fit the door meant cutting away part of the inside of the door. The new door top will eventually be tied to the inner door again so the door top has enough strength.

After doing the driver's door, it's a little easier to get the same effect on the passenger door, shown here.

Though most of the chop is done, you can see where the rear window was cut 1in for aesthetic reasons. The package shelf had to be shortened 2in so the rear seat attachment point could be moved back to its original position.

Right
Another view of the rear of the old Plymouth; seams show how Mike pivoted the rear body section forward to meet the new roof position.

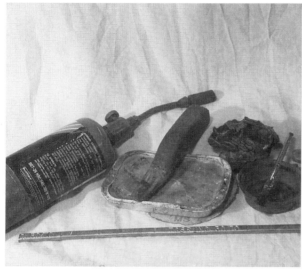

Mike chooses to "lead" the seam at the front of the roof. Necessary tools are shown here: a small torch, beeswax and paddle, steel wool, tinning acid, and the lead.

The process starts by cleaning the seam with a wire brush.

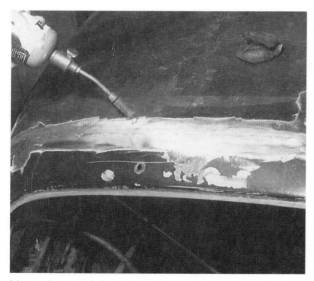

Next, the torch is used to drive any moisture from the seam.

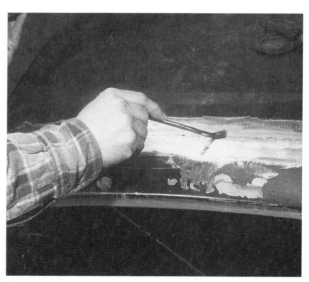

Tinning acid is next—used to further clean the seam.

Sticks of lead are actually closer to solder, made up of 30 percent tin and 70 percent lead. Here, Mike melts the lead and flows it into the seam.

A pad of steel wool and more heat are used to encourage the lead to flow between the two pieces of metal and seal the seam.

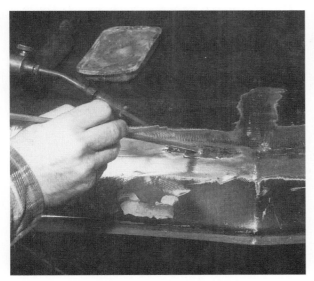

A bit more lead is now melted into the seam...

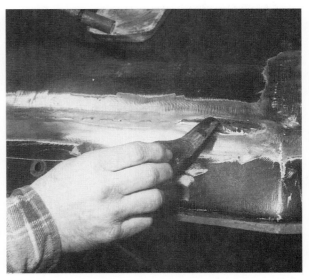

...and smoothed with the paddle dipped in beeswax. Next, Mike will wash the area thoroughly with lacquer thinner to remove any residual beeswax or acid. He does initial smoothing of the seam with a 9in, 24-grit pad on a small grinder (to create a very flexible pad).

The lines of the Plymouth are indeed very natural and pleasing to the eye.

Here's the finished product. Note how well the door matches the rain gutter.

Mike's one-top-chop leaves Pat Barry, the Plymouth's owner, with a nice large windshield to look through.

Chapter Six

1941 Ford Convertible

This 1941 Ford is being chopped at Creative Metalworks by that shop's owner, Kurt Senescall. Some people are afraid to chop a convertible top for fear that the linkages need to be cut—upsetting the geometry and opening up a can of worms. In most cases, however, chopping a convertible top can be accomplished by cutting the windshield posts and then dropping the whole top mechanism down into the car. Of course, it's never quite as easy as it sounds. In the case of the 1941 Ford, Kurt did have to do some cutting of the top mechanism itself (although he reports that on 1946 to 1948 Fords, the top can be dropped into the body intact).

This is the "before" picture, one 1941 Ford Convertible (with some 1946 sheet metal on the front end).

Close-up shows the uncut windshield post and wing window. The grinding marks on the post are from metal repair done prior to the chop.

The first step was, as always, to thoroughly measure the top. Next, Kurt cut down the windshield posts. Then, the folding part of the top was chopped. The individual steps can be broken down as follows:

1. Extend the front crossbow forward until it reaches the lowered front windshield posts.
2. Cut 1 7/8in out of the vertical post (above the main pivot point) so the whole top mechanism drops the same amount. Note that 2in taken out of the posts resulted in 1 7/8in vertical drop. All measurements were taken before the top and windshield posts were cut so that the cut would be even on both sides. Kurt also measured an angle from the back of the door to the top of the windshield posts.
3. The main folding lever has to be cut about 1in in order to get the right angle where the main joint is located above the door opening.
4. The small lever that bolts to the horizontal linkage above the door has to be moved slightly.
5. The rear-most pivot for the top linkage has to be altered so that the top, with its extended header, will fold up into the package tray.

Chopping a convertible is relatively easy if you don't have to disassemble and re-engineer the folding top mechanism. Though each car is different, (as Kurt points out) most convertibles can be chopped by cutting the windshield posts and then lowering the entire mechanism. You might have to find a way to temporarily mount the mechanism at a new lowered location and then work the top up and down until you find and correct any problems that arise (no pun intended) with the new position. The key is to avoid panic. If you absolutely have to make some changes to the mechanism, think your way through each one—it's probably not as complex as you first thought.

For the door glass, Kurt came up with an interesting shortcut. The window frames on the old Ford are made of stainless steel. In order to avoid any unnecessary welding, the lower rail of the window frame is moved up (note the picture), and new glass will be cut. This keeps most of the stock hardware and the stock mechanism (if that's what you want to use) and avoids the need to weld up a chopped frame.

The wing window frame is more difficult. The front channel (parallel to the windshield posts) has to be cut the same amount as the posts (see photos). The piece that is cut out is saved and used to lengthen the short top section, which must also be cut and welded. Kurt welds the stainless channel with the heli-arc and does fuse-welds so there is no filler. If filler is required, the correct number is 304 stainless welding rod (or 308). Kurt notes that aircraft safety wire is actually Number 304 stainless, and he sometimes uses this .032in wire as welding rod because the 1/16in rod is the smallest that's commonly available, and it's too big. The safety wire, on the other hand, is cheap and doesn't leave big puddles of filler like the welding rod might.

Kurt adds a note for anyone working on a 1946 through 1948 Ford, "The window frames on those cars are made of a steel frame with a stainless steel wrap. When you cut these and try to weld them, it is very tough to weld both layers. There's usually rust and impurities between the two layers, and the welding draws out the impurities and it gets very tough to do a nice weld. The trick deal is to strip off the stainless cover, cut and weld the inner steel frame, and then have the frame polished and plated."

Chopping a convertible top isn't impossible or even extremely difficult. In fact, some convertibles are actually easier to chop than similar hardtops.

Rather than cutting right through the center of the posts, Kurt's plan will involve more work but leave fewer visible or hard-to-finish welded seams.

The posts have been cut across the front but not all the way through. The horizontal cut meets a vertical cut on the back side of the post.

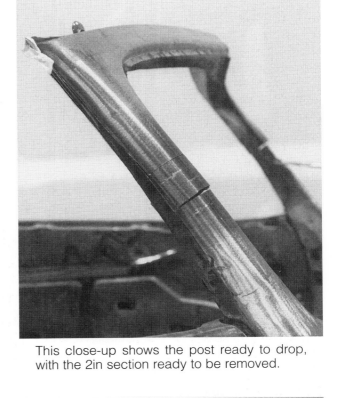

This close-up shows the post ready to drop, with the 2in section ready to be removed.

The chop is done and the real work begins. Note that Kurt left a vertical spear, actually part of the windshield post's base.

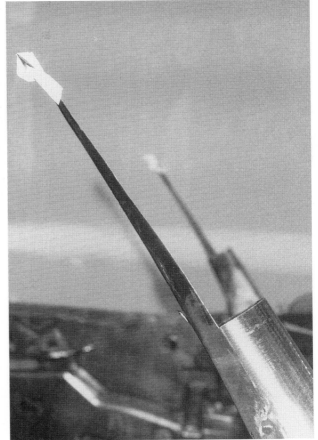

Right
This vertical spear will be used as a guide to line up the cut-down windshield posts, and is also part of a reveal line that Kurt wanted to keep intact.

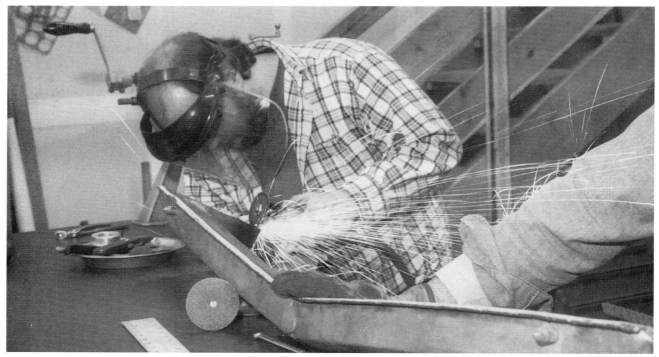

Kurt takes the final cut necessary to remove the 2in section with the cut-off wheel.

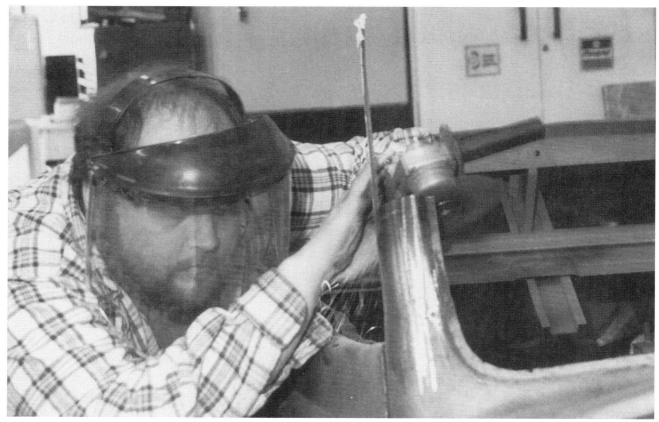

A small grinder is used to carefully clean up the edges before the posts are welded back together.

A trial fit shows the new assembly to be too narrow.

Kurt cuts most of the way through the windshield frame at the corners so it can be spread slightly.

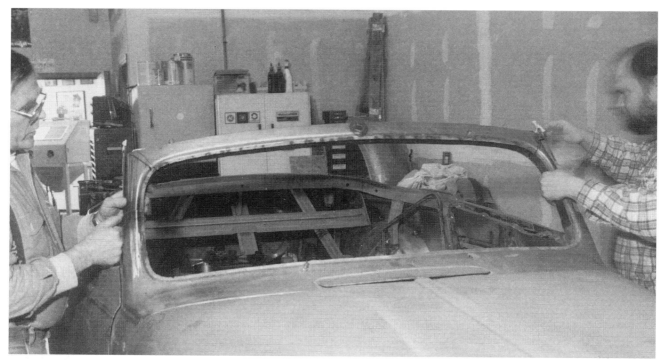

Another trial fit and some careful adjustment.

Close-up shows the slits at the corner.

The windshield frame is clamped in position with a pair of vise grips so more trial fitting can be done.

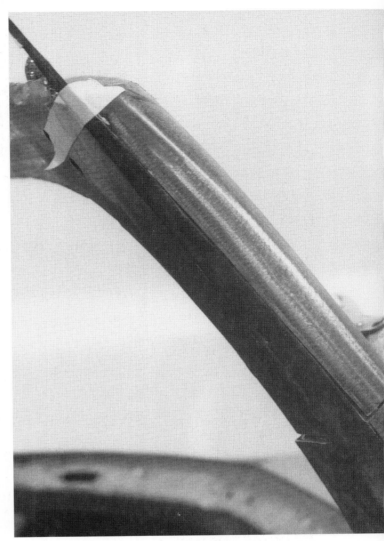

This shot shows the backside of the post and how Kurt neatly slid the post down to the new position, using the spear as a guide and as a way to retain the post's shape.

Trial fit of top mechanism shows it to be both too tall and too short. Front crossbow is extended to make up the missing length.

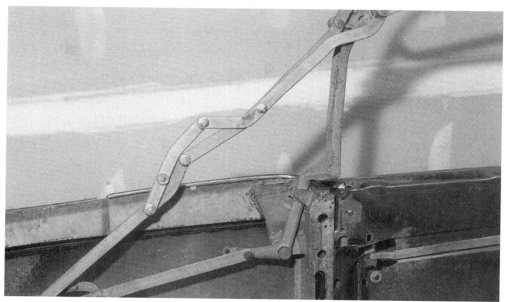

Instead, the vertical bar shown here will be cut 1-7/8in to match the 2in taken out of the front posts.

Though many convertible top mechanisms can be chopped by lowering the whole mechanism into the body, the design of this Ford won't allow for that simple method.

Here we see the top mechanism after the vertical bar has been cut. In addition to the vertical bar, the main folding lever (seen vise-gripped in place here) had to be shortened by about 1in.

Here's the passenger-side post after the welding has been finished.

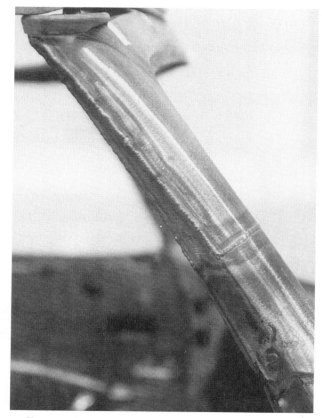

Extremely neat heli-arc welding leaves a barely visible welded seam—the result of skilled welding *and* careful planning.

This shot shows the top mechanism with the vertical bar already cut. Note that the rear-most pivot in the body for the main folding lever was relocated slightly so the top and extended header would all fold up into the package tray correctly.

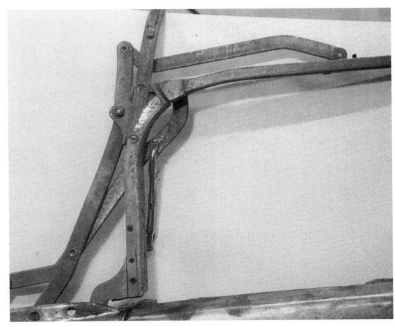

Close-up shows shortened vertical bar and clamped folding arm.

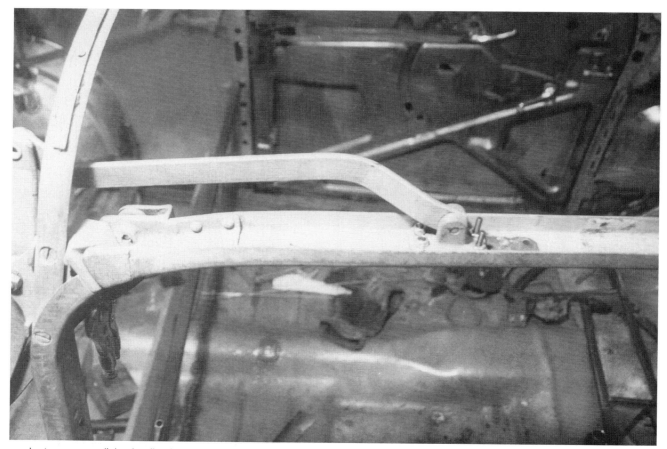

Just one more little detail—the point where this lever, bolted to the horizontal member, had to be moved back slightly.

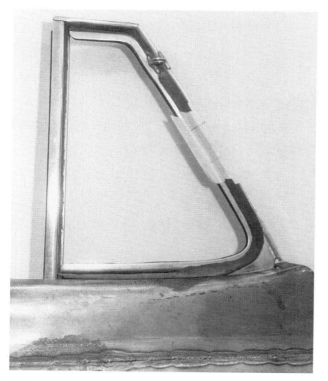

Here we see the stock window frame and, laid over it, the "new" frame.

Left

Wing windows will have to be cut the same amount (shown on the tape) as the windshield posts. The piece that Kurt cuts out will be saved and used to lengthen the short top section of the frame. The frames themselves are stainless, so Kurt will fuse-weld them or use aircraft safety wire (which is actually #304 stainless) as an ultra-thin welding rod to avoid creating big puddles of molten metal.

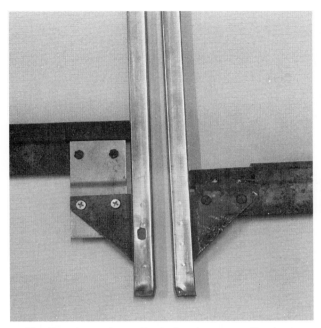

For the door glass, Kurt simply moves the lower rail up and then has new glass cut. This avoids welding and allows use of the stock hardware and mechanism.

Simple stepped bracket is made from mild steel; moves the window up the required amount.

Here's the finished car.

Chapter Seven

1960 Oldsmobile Sedan

Roger Rickey is the motivating force behind Roger's Rod and Customs, a small shop in Anoka, Minnesota. Charlie Julich, owner of the Oldsmobile pictured here, chose Roger to chop the top on the two-door sedan. With more than twenty top chops to his credit and a very good reputation, Roger seemed the ideal person to chop the top on the 1960 Oldsmobile.

What makes this chop unusual isn't just the late-model nature of the car but the fact that Roger did more than just whack the top. He also converted it from a two-door post to a two-door hardtop.

Roger doesn't work from sketches; he started the chop by taking out the front and the center posts, cutting 3in off the rear posts, and sliding the top down and back. Working by eye, Roger just kept dropping the roof farther and farther. Each time he set it farther down, he also moved it farther back. Before deciding on the final position for the top, Roger pushed the car out of the shop so he could get back far enough for a good look.

When he finally got the top in a position where it looked right, it was 3in lower and 4.5in farther back—which in turn left the front of the roof too far back. Instead of splicing in a piece in the center of the roof to move the front of the roof forward, Roger took about 3in from the front of a hardtop roof (the hardtop roof sec-

tion was actually U-shaped and ran along across the front of the existing roof and then back along the sides) and tacked that in place. He tried various positions until it seemed to fit correctly and the front posts lined up.

Where the two roofs meet, along the side of the top, Roger had to cut through the inner support structure in order to join the roofs. Because this inner structure is so important, Roger took pains to reinforce the area and make a good strong roof.

The windows are a big part of this job (they're a big part of this big car). For side glass, Roger cut templates from wood and then sent those off to the glass shop so the glass could be cut to match the templates. The rear quarter windows don't go up and down, but slide in and out. The windshield proved to be the Achilles heel of the whole project. Though Roger planned to trim material from the bottom of an original windshield, he ended up ordering Plexiglas front and rear windows and then trimming the new glass to fit the opening.

Each top chop is different, and this Oldsmobile was definitely unique. In chopping this top, Roger had to do more than just lower the lid. The work involved blending two roofs, creating a new means of locating the rear side windows, and determining the best way to deal with cutting the curved front and rear glass.

Interview
With Roger Rickey

The Minneapolis-St. Paul area is blessed with a tremendous amount of hot rodding activity. When local street rodders discuss top chopping or quality sheet metal work, the name that comes up again and again is Roger Rickey. Roger has chopped a lot of tops; he's a man with a wealth of tips and ideas for first-time top choppers.

Roger, give me some background on yourself: How did you get involved in body work and top chopping?

I started out going to school right here at Anoka Tech. I went for one year, to the body and paint class. It was the basics, a little of everything. And from there I started working my way from body shop to body shop, trying to better myself each time I moved. Then this friend of mine brought me this guy's 1949 Mercury for some body work. That was how I got started working on old cars. But I was still working at body shops in the day. More and more work came into my home shop, and I would have to do that work at night, after putting in a full day at the other job.

One day I thought, I should open my own business. I had acquired tools over the years and I had an air compressor for my small shop at home. I thought, hell, I've got nothing to lose, and I started my own business on July 1, 1986. I figured if it didn't work out I could always go back to working in a conventional body shop—but I haven't had to. I've been on my own ever since.

Actually, when I started on that Mercury in 1982 I started working on my metal-working skills, and that's what really helped. I bought a book on metal working and I've tried to get better year by year. Now when I work, I always work from a cardboard template. And I've gotten better at shrinking and stretching.

How did you get into chopping tops?

I don't know, I like doing it. I like doing things I haven't done before, building a car I haven't built

Roger Rickey attributes his success with top chops to his above-average metal-working skills and his willingness to tackle new projects.

before. I just felt that I like doing things that aren't ordinary—and that there's nothing I couldn't do. If it's making something out of three pieces or ten pieces, I can do it.

Would you say metal-working skills are essential to doing a good top chop?

Yes, definitely. I've seen a lot of tops that other people have done. I don't know why, but people either leave things out or they take a shortcut. Maybe that's why you see glass breakage on some of these cars. Especially those cars with curved glass.

When people screw up a top chop, can you say there's one or two major reasons they got into trouble?

I think people have a tendency to "quarter-section" a roof—you know, cut it up into four pieces. I think that's where they really get lost. Because then they've got it in four pieces and they say, 'Where do we go from here?' I've never ever quartered a roof and I've made them all work. People do too much cutting.

The other area they screw up on is the rear sail section between the quarter window and the rear glass. There's an area right there on the corner of the roof that people can't figure out. They don't realize they have to relieve the area, and then if they do, they either put too little metal in there or too much, and then they have a big pucker and they don't know how to get it straightened out.

Some cars are tough, like the '39 and '40 Ford sedans. When they get the top done, the lines aren't right. That's like an aero sedan, a Chevy or Cadillac. If the line doesn't come down right, they look really odd. Those are the things that come to mind when I think of mistakes.

Any time you have to get in and cut near the center of the roof you can really get in a lot of trouble. As you move toward the center of the roof, you get more flop. And the flatter the roof is, the harder it is to cut because it doesn't have as much strength. There's a lot of strength in the shape.

This is the 1960 Oldsmobile at Roger Rickey's shop in suburban Minneapolis. This shot, from early in the project, shows the lowered top, tack-welded into position.

I like to cut pretty close to the front. When I do a splice to lengthen the roof, it's always at the front. They're usually pretty domed at the front, and there's more strength there. There's less problems when you do it there. When I splice a roof, I overlap it at the front where the splice is. I overlap it an inch or so, then tack-weld it all together until you've got the right shape. Then you cut through the bottom layer of metal, right where the two pieces meet. Then butt-weld the two pieces together as you're cutting and trimming the bottom piece away. But the farther away you stay from the center, the better off you are.

Any roof I've done, the splices in the posts never match up. So you're always cutting and then pulling something out to make it the right width. I've done a lot of roofs where I cut one side and didn't have to do that to the other side. A lot of those old cars, the tolerances aren't that close.

When you do a top chop, how do you plan it out?

Mostly by experience. How much do I want to cut; what are the features the customer wants? Will there be rain gutters, for example? How much lean will there be to the windshield?

People want to know how much to chop a top, but they haven't thought about all the features they want and all the little details. These details make a lot of difference and affect how much work the project turns out to be.

The strength of the floor pan has a lot to do with the top. If the floor is weak, then don't even chop it. Fix the structure first (I don't like to see people cut the whole floor pan out; it lets the body move around too much). You've got to have a foundation for the whole car. The doors must align. The hinges must be solid. The doors must be solid and fit correctly. All that must be in place before starting on the top chop. The preparation is very, very important. People should do their rust and structural repairs first, before they start the chop.

Do you do pictures or sketches before you start?

No, most people go see a car at the show and then they say, 'I want it chopped just like this other car I saw.'

You work by eye a lot?

Yes, people should get the car out of the shop once in, while during the work and have a look at

Roger cut the front windshield posts off almost flush with the body and left about 3in of the post attached to the hardtop front roof section.

it. It's hard to do in Minnesota in the winter, but it really helps. A lot of times it can look OK in the shop, but when you get forty feet away, you can see a problem more easily.

They say you should measure, but sometimes the original dimensions are way off from one side to the other, so sometimes it comes down to how it looks.

Are you careful to brace the car before you take the top off?

Bracing it before you start is very, very important. That Olds is all braced. If you don't brace it up and keep the panels in the position they're in now, you're going to do a lot of work to get the panels back in position after you cut the top off. Because your cowl will move forward or back after you cut the top if it isn't braced, and that throws the door off. It's like a tin can: when you cut the top off, the sides get flimsy as hell. You've got to keep everything together until you're ready.

I chop the top and do all the welding on the posts and the back window, and then when I overlap the roof at the front, I tack the windshield area last.

How about some hints for working with glass?

You can make a one-piece window (eliminate the wing window) pretty easily. You have to make all the channels and get the fuzzy liners and cat's whiskers in there, get all that done first. Then make a paper pattern from .020in board. I call it "potato-chip-box cardboard," maybe a little thicker. Make a pattern, make it go up and down in the channel. Then transfer those dimensions to a piece of 0.25in board. Make sure that board fits just right and goes up and down right. Countersink all the screws in the channels, no rivets. Get the correct little taper-headed screws—they're really small, and the head sinks into those black fuzzy channels. If the board goes up and down right, then take it to an auto glass place and have them cut the window to exactly that dimension.

The new front section is actually U-shaped and wraps around to the sides of the roof. Roger had to cut through the roof's inner structure when he joined the two roofs. Here we see how he cuts a slot in the sedan roof to provide access to the inner structure for reinforcement.

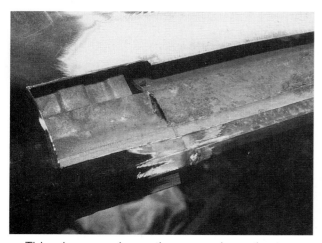

This close-up shows the area where the two inner structures meet.

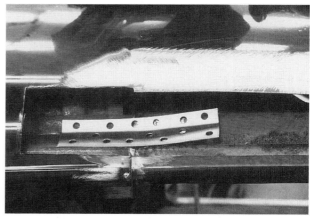

Roger makes a reinforcing plate and punches it full of holes so he can weld through them and not just at the edges of the reinforcing plate.

If people would do that much, they would avoid a whole lot of trouble. The cat's whiskers and fuzzy channels are available; I get them at the National Chevy Association [see address in the appendix]. They've got all the channels—flexible channels, all that stuff. Always screw the fuzzy liners in; the little screws have to come from a good hardware store. Make sure the screw sinks into that fuzzy liner. If you leave one head sticking out

just a little, the glass will find it, I guarantee it.

For a glue-in window, make the pattern yourself to be sure it's done correctly. Make the pattern small enough so it doesn't touch the metal at the outside of the opening (never let the glass touch any metal). When you get the glass, use a windshield ribbon kit. Cut the ribbon, put the glass in, and pat it down. You can see where the ribbon made contact, then use black urethane

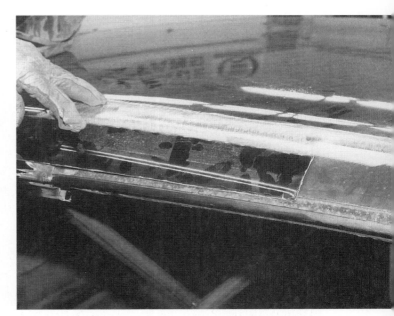

The wire-feed welder is used to weld the plate. This will help give the roof—and the entire body structure—plenty of strength. When Roger cuts through a windshield post with inner bracing, he follows a similar procedure to make the post as strong as possible.

Roger takes a nicely crowned piece of metal from the sedan roof to repair the body skin after the inner structure is repaired.

adhesive-for-glass in a caulk gun (tape off the paint, because it makes a heck of a mess), then smooth it off so it's flush and neat. Silicone is messier; it yellows and attracts dust. The urethane sets up and makes a neater job.

When do you need tempered glass, and when should it be laminated?

Any door glass, if you don't use the original channel that surrounded the glass, needs to be tempered. Now if you want to eliminate that surrounding channel, you have to cut the glass according to your pattern, and then have it tempered. Of course, you need to cut and sand the glass first, then have it tempered.

All the newer cars, like hardtops, with no surrounding channel, that's all tempered glass. So laminated glass works in most situations, but not for a freestanding hardtop window with no stainless steel trim around the edge.

You can have curved glass made, by making a complete curved template of the glass done in aluminum sheet, for example. I've seen the ads in *Hemmings Motor News*, but it's real expensive, so I've never done it.

How do you cut a curved windshield?

For the Olds, I'm going to sandblast the windshield. When you sandblast with sand, you can see the friction, a red glow at the edge, so when you blast glass, you have to keep moving so you don't cause it to crack due to the heat.

Any more words of wisdom for the first-time top chopper?

Don't be afraid to ask. You're going to make mistakes—ask some questions. Before you cut the roof, make sure you know how much you want to go. Be sure the doors fit, make sure it's bolted to the frame. And double-check all the time to be sure the sides are even, be sure to brace everything up real good.

What about trim around windows?

Well, you almost always lengthen the top, so you usually don't have enough garnish moldings. If you put wing windows back in the car, go find another pair of wings and door tops and garnish moldings. Why fabricate the 3 or 4in piece if you don't have to? It's just like the top itself; if you need to splice in a 4in piece, why not use the front of another roof, instead of a 4in piece of sheet steel? That way, you only have to weld up one seam in the roof, not two.

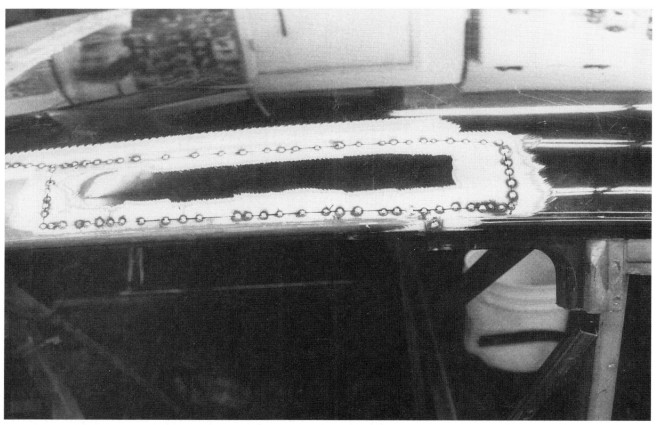

The patch on the left side is cut to exactly fill the hole, then tack-welded into position.

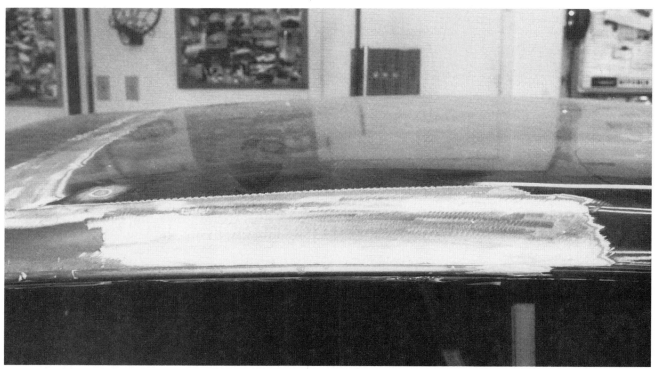

This shot is out of sequence, but it shows how neat the patch turned out after finish-welding and sanding.

Rear three-quarter view shows the Oldsmobile with the new roof tack-welded and clamped into place. Essentially, Roger dropped the sedan roof 3in and slid it back almost 5in. The front section of a hardtop roof was used to make up the missing material.

Vise grips are used to hold the rear posts in place. How to deal with the rear quarter windows was only one of the dilemmas Roger had to face.

Note how far back the old upper door post is relative to the back of the door; that's how far back Roger pushed the sedan roof.

Roger has cut out the remains of the old sedan post and the trim for the rear window.

Rear window posts have been welded into place. The new window hole is smaller than the original, which will make it hard to create the new "glass." The original rear window is tempered glass—which can't be cut the way laminated glass can.

The rear side glass will slide in and out of new channels that haven't been installed yet—eliminating the need for window-crank assemblies.

The rear posts are butt-welded after the top is in the correct position.

After he has the two tops positioned where he wants them, Roger splices in the rest of the front wind-shield post. Though the splice looks like one piece, Roger actually made the splice from two pieces—one cut from the sedan post, and one from the hardtop post.

The windshield retains the shape of the original roof. Roger's initial plan was to trim the original windshield at the bottom to make it fit the new smaller hole. In the end, a new Plexiglas windshield was formed by a local glass company.

Because the hardtop wing windows are surrounded by cast channels that can't be cut and welded, these sedan wing window assemblies will be cut and welded.

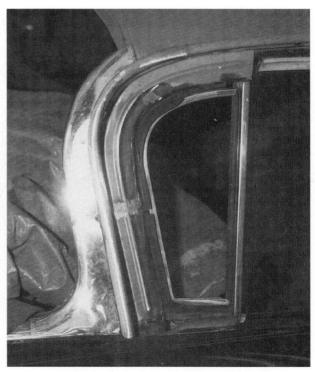

Roger carefully cut and welded up the wing window frames to fit the new roof's dimensions.

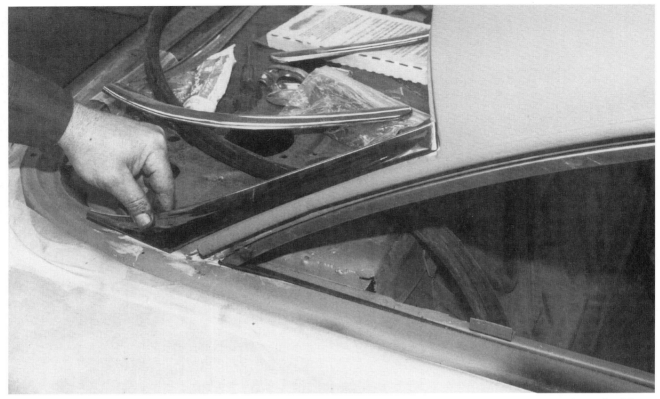

Garnish moldings and trim can be very tough parts of the small puzzles that make up a top chop. In this case, Roger could simply trim an existing trim piece after the top and/or window opening gets longer, necessitating the creation of longer pieces of trim and molding.

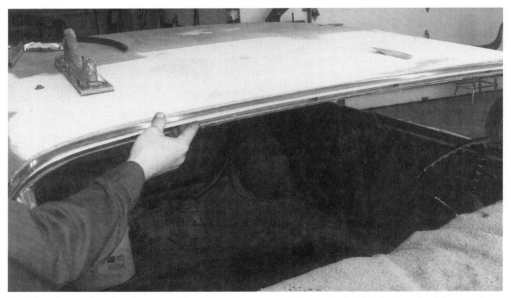

Trim from the hardtop roof will work around the windshield. For inside work where the opening is bigger than stock, Roger recommends using additional pieces of trim from a parts car and simply splicing and cutting two pieces of trim to fit.

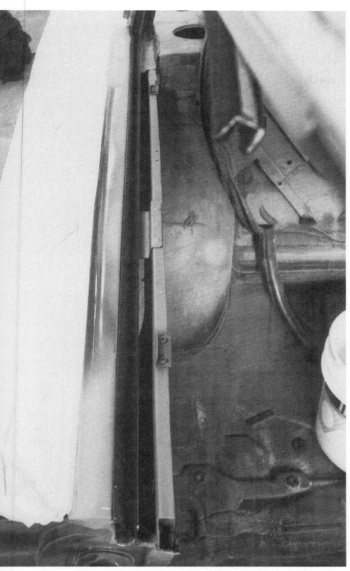

Upper channel for the rear windows is U-shaped mild steel, bent at an outside shop to match the inside radius of the rear window opening. These channels were welded in after the reinforcing square tubing was welded up into the roof, above the area where the channels would be mounted, to give the roof more strength.

Bottom channel of rear side glass is actually stainless steel channel, with fuzzy liner installed. Channel is tipped down slightly at the front so water will run out.

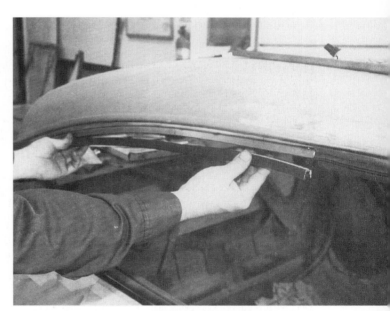

After the channels are installed, Roger will install the liners with screws (never use glue).

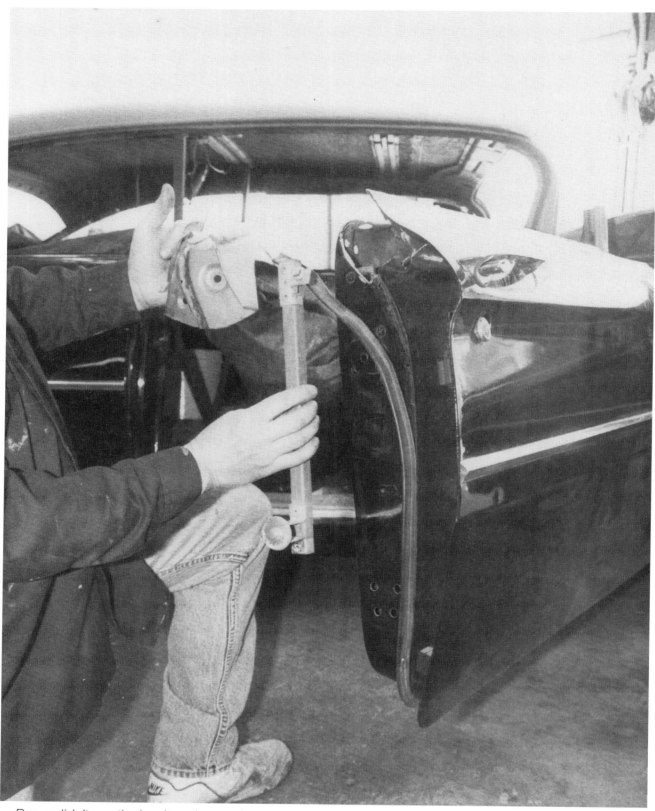

Roger didn't use the hardtop doors—they were too rusty. Because the door skin is the same for both, he decided instead to take the rear channel and the upper, inner door section (held in his left hand) from the hardtop doors and use them to convert the sedan doors.

New hardtop rubber strip is used at the top of the roof, where the door glass will roll up against it.

Here's the finished Olds.

Note how the chopped roof lines of the completed Olds complement and accentuate the look of the entire car. A successful chop does more than lower the roof; it redefines or transforms the entire car.

From the rear, the curved glass that makes late-model cars more difficult to chop is evident.

Chapter Eight

1966 Dodge Charger

The 1966 Charger seen here is owned by Bob Larsen, from suburban Minneapolis. Though Bob worked for many years as a mechanic, this is his first top chop. Bob's idea was to give the Charger a mild chop by cutting a relatively small amount of metal from the front windshield posts, leaning the posts back to meet the lowered lid, and simply slitting the rear of the fastback roof so it will act as a hinge and allow the top to drop. Bob explained that he always liked the looks of the

This is the before picture of Bob Larsen's 1966 Dodge Charger. Note that the glass and interior have been stripped from the car.

To support and lift the front of the roof, Bob made this simple fixture and added a small screw jack.

Bob and his friend Pat Meger position the jack up against a brace on the inside of the roof.

At the back of the car, Bob lays out a Z-cut that will be followed with the saw.

Bob starts the cut with a cut-off wheel and decides that a Sawzall would really be the ideal tool because it will cut both the outer skin and inner structure in one pass.

Chargers but thought the car would benefit by having the roof dropped and the windshield laid back, to give the car a more sleek silhouette.

Bob started the project by pulling out all the glass and the interior of the old Dodge. Bob also took the precaution of bracing the body opening before the top was cut loose. By starting with a black-and-white drawing of the car and chopping the drawing, Bob was able to determine how much he wanted to chop the top.

One of his first tasks, after stripping the inside of the car, was to make an adjustable structure to support the front of the roof and aid in lowering it. The simple device was made from rectangular tubing and an old screw jack. At the back of the roof, Bob simply laid a two-by-four under the roof, supported by another screw jack.

One of the hardest parts for Bob was trying to decide how to deal with the inner roof structure at the back of the roof. Though he eventually took a pie-cut out of the inner structure, it might have been easier to cut out nearly all of the inner struc-

ture and then reinstall it after the top was lowered and all the seams were tack-welded into position.

Bob intended to do the cutting with a cut-off wheel, either on an air-powered zinger or a 4in electric grinder. But part way into the job, he and his friend Pat Meger decided a Sawzall would be essential, partly because the long blade would allow them to cut through the inner and outer structures at the same time.

At the back of the roof, Bob did a Z-cut so one panel would slide by the other as the top came down. A piece, 1.5in long, was cut out of the windshield post at the top, then the bottom of the post was pie-cut and leaned back to meet the new position of the roof.

It all sounds perhaps simpler than it really was. Yes, there's still finish work to do, but the point is that Bob did his own chop with just the help of a friend, a few basic tools, and a wire-feed welder. What helped is the planning that Bob put into the project ahead of time—which kept the whole thing relatively simple.

138

Here's the Z-cut, which will be further modified before the top comes down.

Sawzall in hand, Bob finishes the Z-cut at the back of the car. The idea is to do a modest chop by hinging the top at the very back of the car and letting it come down in the front.

Because Bob knew how much the top would come down, he was able to snap a "before" and "after" line with a chalk line, thus marking the pie-cut he would need to take on the rear side panel.

This shot shows the pie-cut that was taken at the rear—from the inside of the car.

Here's the old Dodge with the pie-cuts taken out and the short sections removed from the front posts.

The Sawzall is a nice neat way to take two short 1.5in sections from the front posts.

Rear view shows how the original Z-cut was modified and brought much farther back before the top was chopped.

The rear of the roof is supported by another simple screw jack and a pair of two-by-fours.

The front post after the 1.5in section is removed and before the top is dropped.

After the drop, this is the new profile.

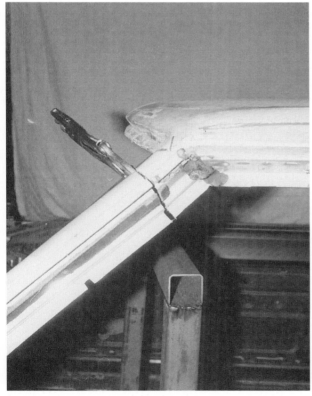

By taking a pie-cut at the bottom of the front post, Bob was able to simply bend the post back to meet the new roof position.

A template of the new profile, made up before the top was chopped, is set into place as a way to test the top chop and to make sure it's even from side to side.

Rear three-quarter view shows very pleasing lines and some interference at the back Z-cut.

143

Here, the old original windshield is laid in place as a guide to indicate how much material will have to be removed to make a good windshield.

Z-cut area, after the initial welding and some sanding.

Here we see how the inner structure was welded up after the top was in final position.

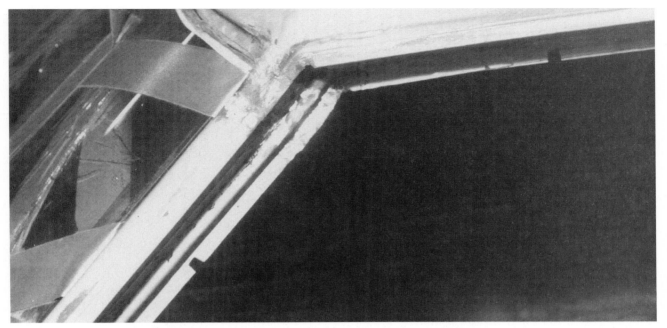

Front post is welded up and sanded. Old windshield has been rough-cut to fit.

Bob's Charger, with the roof in its final position and the initial welding done.

This angle shows off the new Charger with its sleek silhouette and rakish windshield angle.

Outside at last, this subtle chop is hardly noticeable.

Z-cuts and welding seams have disappeared under a single coat of body filler.

Chapter Nine

A Group of Coupes

1948 Ford Coupe

This five-passenger coupe belongs to Kevin Lehan, from suburban Minneapolis. Though he's a Harley guy through and through, Kevin also enjoys four-wheeled hot rods. The four-wheeler in Kevin's garage is a 1948 Ford coupe, and Kevin is quick to explain that he did the top chop himself. The vertical drop was 2.5in. Kevin didn't want to "cut the top into many pieces," as he puts it, so he planned out

Kevin Lehan didn't want to cut the top of his 1947 Ford into many pieces and chose instead to cut it, move it forward, and change the angle of the rear window.

As with the other Ford coupe in this chapter, Kevin cut the rear window out as a unit before chopping the top.

a top chop that would keep the top in one piece. Before starting on the chop, he cut the rear window out as a complete unit. With that done, Kevin went on to cut the front posts, the center post, and the rear of the roof. When the top came down 2.5in, it also went forward 1.5in. Kevin chose to keep the stock windshield angle, though the top of the windshield posts had to be modified to meet the roof.

Because the top moved forward, the top of the center post moved too far forward and had to be cut out of the top and moved back to line up with the door. The rear window was laid down at a steeper angle than before (the final angle was determined by eye), but a gap remained between the bottom of the window and the top of the rear deck lid—one which will need to be filled.

Kevin will also have to fill and finish the area between the back of the quarter windows and the metal surrounding the rear windows. In spite of the filling that will be needed, Kevin avoided cutting the top into two or more pieces and having to do all that welding and finish work. When he's all done, Kevin will have a very slick old coupe to park alongside his Harley-Davidson.

Here we see the rear window clamped in place. The window's final angle and position were determined by eye.

The vertical drop is 2.5in. Windshield angle is stock.

Here you can see how everything moved when the top came down.

1939 Ford Coupe

Dan Burkholz, of suburban Minneapolis, wanted more than a standard top chop for his 1939 Ford coupe. Instead of the standard 2 or 3in top, Dan decided to chop the top and create a hardtop at the same time.

As a somewhat unusual planning step, Dan not only drew a sketch of the new car but also built a scale model of his car, including the top chop and the elimination of the center post.

And while some builders take off the rain gutters, Dan decided to leave his—because the cars look too "bald" without them.

Dan cut the rear window out of the top, then cut the top at the front posts and about halfway up the back part of the rear window. Next, the lower part of the rear window and the section from the door opening back were cut away from the car. Then, when the top was dropped, Dan moved the rear corner of the roof forward to meet the new roof position. The rear window was clamped into place (forward from its original position), leaving a gap in the rear deck between the bottom of the new window position and the top of the deck lid.

Dan wanted a short-top look and liked the idea of a longer deck lid. One of the things he doesn't like about many of the 1939 and 1940 top chops he sees is the way the rear quarter window gets small and pointy when the posts are cut at the bottom and the top is then dropped two or more inches. By cutting the top halfway up the quarter window, Dan will keep the nice curve at the back of the window. By cutting out the door post, he will enhance the lines of the coupe.

As an aid in cutting the roof, Dan made a small template for the posts where they would be cut. Because of the template, he was able to easily figure how much of the post to remove in order to achieve a 2.5in vertical drop.

Dan was careful to make sure the doors fit correctly, that the floor pan was in good condition, and that the body was securely bolted to the frame before starting any cutting. He also took pains to reinforce the body opening before he cut the top off.

Left
In particular, the top went forward, which meant the center post had to be cut out and moved back to line up with the door.

Dan cut the rear window out of the top before starting on the chop.

Here's Dan's '39 Ford coupe midway through construction. Dan dropped the top 2.5in vertically, bringing it forward as well as down.

After cutting out the rear window, Dan cut the top at the windshield posts and the rear window posts.

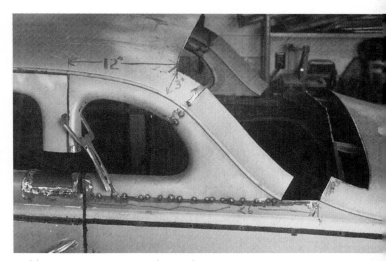

Here you can see where the rear cuts were made and how the lower part of the rear quarter window was moved forward to match up with the new top's position. Dan was careful to cut the top in such a way that he retained the pleasant rounded shape of the rear quarter window.

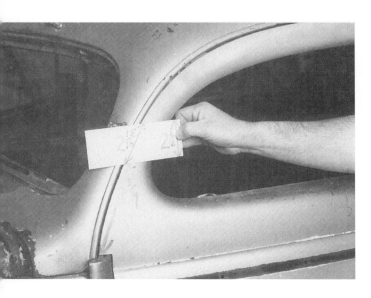

Left
As part of Dan's planning work, he made this template so he would know exactly how much to take out of the post in order to achieve the 2.5in drop.

Like all good builders, Dan made sure he had a solid floor pan before starting and then took pains to reinforce the body before the top was cut off.

More planning. Dan went so far as to make a model of his coupe with the top chop and elimination of the door post.

This mock-up piece taped to the top helps Dan visualize what the top will look like without the center post.

This porta-band saw was used for most of the top cuts. Like a Sawzall, it will cut through the entire post in one pass, and it does so with more precision because both ends of the blade are controlled.

With the rear window clamped in position, you can see how far forward the top was moved and how much material must be added between the rear deck and the rear window.

1941 Plymouth Coupe

The chop on this 1941 Plymouth coupe was performed by Roger Rickey of Anoka, Minnesota. Roger and the owner decided on a 3.5in chop and a stock windshield angle. To help give the car a smooth look, Roger took off the rain gutters. Because he likes a little additional rake on most street rods, Roger dropped the front of the top a little more (a total of 4in) than the back.

The Plymouth retained the stock windshield and rear glass angles; thus, as the top came down, it also became too short. Roger solved that problem by welding in a thin strip of sheet metal (about 4in wide) across the top. The additional strip of metal was cut from a sheet of cold-rolled, eighteen-gauge steel, and welded in as far forward on the roof as possible.

By chopping the top, Roger helped to transform a lumpy old Mopar into a modern, smooth, street rod.

The before picture of the too-tall 1941 Plymouth coupe. *Roger Rickey*

This is a good look at the coupe before Roger drops the top. *Roger Rickey*

Dropping the top meant adding material to lengthen it. Here you can see where the strip of steel was added to the Plymouth's roof.

Right, the finish work on the rear is immaculate and requires minimal body filler.

By taking a little more from the front of the roof than the rear, Roger created a rake that adds to the car's lines.

The front of the roof came down 4in. Note how nicely the door fits the opening.

A look at the inside of the door shows the splices where Roger cut and pieced the new door top together.

Old Mopars aren't known for their great looks, but this ugly duckling shed almost four inches of top height to become a sharp, stylish custom.

Sources

Allied Plastics
Manufacturers of Custom Plastic Rear Windows
3005 North Ranch View Ln.
Plymouth, MN 55447
(612) 553-7771

Roger's Rod & Customs
Roger Rickey
1851 Sims Rd. NW
Oak Grove, MN 55011
(612) 753-4393

Car Creations
Mike Stivers
9199 Davenport St. N.E.
Blaine, MN 55449
(612) 784-0298

Creative Metalworks
Kurt Senescall
9206 Isanti St. N.E.
Blaine, MN 55449
(612) 784-2997

Cory Harder
P.O. Box 337
Mountain Lake, MN 56159

The Eastwood Company
580 Lancaster Avenue
Malvern, PA 19355
800-345-1178

Hot Rods by Boyd
10541 Ashdale
Stanton, CA 90680
(714) 952-0700

Harmon Glass
2001 County Rd. C
Roseville, MN 55113
(612) 636-4676

National Chevy Association
947 Arcade
St. Paul, MN 51060
(612) 778-9522

Metal Fab
Jim Petrykowski
1453 91st Ave. NE
Blaine, MN 55434
(612) 786-8910

Doug Thompson
8204 Laurel
Raytown, MO 64138

Index